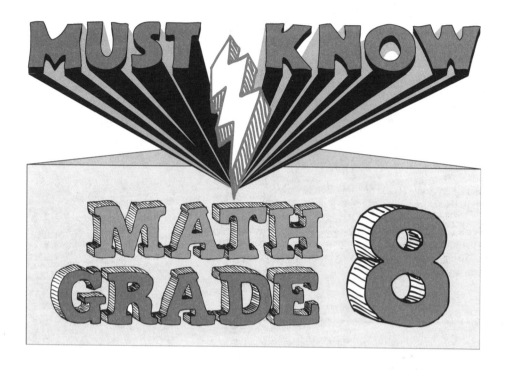

MUST KNOW
MATH GRADE 8

Nicholas Falletta

Mc
Graw
Hill

New York   Chicago   San Francisco   Athens   London   Madrid
Mexico City   Milan   New Delhi   Singapore   Sydney   Toronto

1 2 3 4 5 6 7 8 9 LCR 25 24 23 22 21 20

ISBN   978-1-260-46802-1
MHID   1-260-46802-X

e-ISBN   978-1-260-46803-8
e-MHID   1-260-46803-8

Interior design by Steve Straus of Think Book Works.
Cover and letter art by Kate Rutter.

McGraw-Hill Education books are available at special quantity discounts to use as premiums and sales promotions or for use in corporate training programs. To contact a representative, please visit the Contact Us pages at www.mhprofessional.com.

In memory of Garret Lemoi, my remarkable editor whose unfailing skills and personal warmth helped make this and several of my other books so very much better than I ever hoped they could be.

Contents

Introduction

Welcome to your new math book! Let us try to explain why we believe you've made the right choice. You've probably had your fill of books asking you to memorize lots of terms (such as in school). This book isn't going to do that—although you're welcome to memorize anything you take an interest in. You may also have found that a lot of books make a lot of promises about all the things you'll be able to accomplish by the time you reach the end of a given chapter. In the process, those books can make you feel as though you missed out on the building blocks that you actually need to master those goals.

With *Must Know Math Grade 8,* we've taken a different approach. When you start a new chapter, right off the bat you will see one or more **must know** ideas. These are the essential concepts behind what you are going to study, and they will form the foundation of what you will learn throughout the chapter. With these **must know** ideas, you will have what you need to hold it together as you study, and they will be your guide as you make your way through each chapter.

To build on this foundation, you will find easy-to-follow discussions of the topic at hand, accompanied by comprehensive examples that show you how to apply what you're learning to solving typical 8th-grade math questions. Each chapter ends with review questions—more than 250 throughout the book—designed to instill confidence as you practice your new skills.

This book has other features that will help you on this math journey of yours. It has a number of sidebars that will either provide helpful information or just serve as a quick break from your studies. The **BTW** sidebars ("by the way") point out important information, as well as tell you what to be careful about math-wise. Every once in a while, an 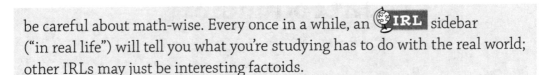 IRL sidebar ("in real life") will tell you what you're studying has to do with the real world; other IRLs may just be interesting factoids.

In addition, this book is accompanied by a flashcard app that will give you the ability to test yourself at any time. The app includes 100-plus "flashcards" with a review question on one "side" and the answer on the other. You can either work through the flashcards by themselves or use them alongside the book. To find out where to get the app and how to use it, go to the next section, "The Flashcard App."

We also wanted to introduce you to your guide throughout this book. Nicholas Falletta is a veteran education writer and the author of McGraw Hill's *SSAT/ISEE*. We're glad to have the chance to work with him again. Nick has a clear idea what you should get out of a math class in 8th grade and has developed strategies to help you get there. He's also seen the kinds of pitfalls that students can fall into and is an experienced hand at solving those difficulties. In this book, Nick applies that experience both to showing you the most effective way to learn a given concept and how to extricate yourself from any trouble you may have gotten into. He will be a trustworthy guide as you expand your math knowledge and develop new skills.

Before we leave you to the author's sure-footed guidance, let us give you one piece of advice. While we know that saying something "is the *worst*" is a cliché, if anything *is* the worst in the math you'll cover in this grade, it could be working with functions. Let Nick introduce you to functions and show you how to work confidently with them. Take our word for it: learning how to handle functions will leave you in good shape for the rest of your math career—and in the real world, too.

Good luck with your studies!

The Editors at McGraw Hill

The Flashcard App

This book features a bonus flashcard app. It will help you test yourself on what you've learned as you make your way through the book (or in and out). It includes 100-plus "flashcards," both "front" and "back." It gives you two options for using it. You can jump right into the app and start from any point that you want. Or you can take advantage of the handy QR codes at the end of each chapter in the book; they will take you directly to the flashcards related to what you're studying at the moment.

To take advantage of this bonus feature, follow these easy steps:

Search for **McGraw Hill Must Know** App from either Google Play or the App Store.

↓

Download the app to your smartphone or tablet.

↓

Once you've got the app,
you can use it in either of two ways.

↙   ↘

Just open the app and you're ready to go.	Use your phone's QR Code reader to scan any of the book's QR codes.
You can start at the beginning, or select any of the chapters listed.	You'll be taken directly to the flashcards that match your chapter of choice.

↘   ↙

Get ready to test your math knowledge!

Author's Note

Must Know Math Grade 8 builds on the skills and concepts that you learned last year. Examples of clear-cut problems with complete solutions are provided in all the basic area of mathematics you'll need for success in this school year.

Mathematics in general, and this book in particular, are not meant for you to read straight through. This book contains plenty of example problems with complete explanations of their solutions that will guide you through the steps involved. Always try to do the problems on your own before you read the explanations. This process will give you a reasonable benchmark of how much you already understand, and it will clarify what you are having problems understanding.

After trying an example problem, check the solution. Even if you got the correct answer, look to see that you solved the problem in the best and most efficient way. If you have stumbled getting the right answer, remember that's why you bought this book. Read through the explanation and then close the book and try to solve the problem again.

At the end of each chapter, you'll find plenty of new problems in the Exercises section. These problems will make sure you consolidate and apply your understanding. Once you work through a set of problems, check the Answer Key at the back of the book. You'll be pleasantly surprised because it gives you more than just the right answers. For each problem, you'll find a detailed solution based on the model explanations offered in the chapter examples. The book also offers more than 100 electronic flashcards tied to all the lessons that you can access online.

Each chapter of the book begins with statements about what you **must know** when you have completed the chapter. After working through a chapter, if you don't feel that you have a good grasp of the concepts described at the beginning, make a point of going back through the chapter until you do. The more you are willing to do this, the more skilled and confident you'll become.

Real Numbers

MUST KNOW

⚡ The base-10 number system gives value to a number through the position of the digits 1 through 9—and 0!—relative to a decimal point.

⚡ Integers are whole numbers that are either positive (1, 2, 3, etc.), negative (−1, −2, −3, etc.), or 0.

⚡ Rational numbers can be expressed as the quotient of two integers (a fraction), while irrational numbers cannot be expressed as a quotient. Together, they form the real number system.

⚡ The absolute value of a number is its distance from 0 on a number line.

⚡ The properties of numbers explain how real numbers work together when performing addition and/or multiplication.

1

he base-10 number system is thousands of years old, and it is used everywhere today. This system gets its name from the fact that it uses ten digits to form all numbers: 0, 1, 2, 3, 4, 5, 6, 7, 8, and 9. With just these few digits, every real number can be represented—from 1, 2, and 3 to infinity!

The Base-10 Number System

In the base-10 system, the value of a digit depends on the place it occupies in the number. For example, in the number 235, the 2 represents 200 because it occupies the hundreds place, and 2 times 100 equals 200. In 325, the 2 represents 20 because it is in the tens place, and 2 times 10 equals 20. Can you figure out what the 2 means in 124,356? To do so, it's helpful to review the names of the places in a place value chart like the one shown below.

Billions	Hundred Millions	Ten Millions	Millions	Hundred Thousands	Ten Thousands	Thousands	Hundreds	Tens	Ones

EXAMPLE

▶ What is the value of 2 in the number 124,356?

▶ Write the number in a place value chart like the one shown above.

Billions	Hundred Millions	Ten Millions	Millions	Hundred Thousands	Ten Thousands	Thousands	Hundreds	Tens	Ones
				1	2	4	3	5	6

▶ Identify the place name that corresponds with the digit 2. The digit 2 appears in the ten thousands place. Therefore, 2 times 10,000 equals 20,000.

▶ The 2 in 124,356 is in the ten thousands place and represents 20,000.

Let's try another example.

EXAMPLE

▶ What is the value of the digit 7 in 3,718,432?

▶ Write the number 3,718,432 in the place value chart.

Billions	Hundred Millions	Ten Millions	Millions	Hundred Thousands	Ten Thousands	Thousands	Hundreds	Tens	Ones
		3	7	1	8	4	3	2

▶ Identify the place name that corresponds with the digit 7. The digit 7 appears in the hundred thousands place. Therefore, 7 times 100,000 equals 700,000,

▶ The 7 in 3,718,432 is in the hundred thousands place and represents 700,000.

Just as there are names for the places in a whole number, there are names for the decimal places. Using a place value chart can help us identify the value of a digit when the number is a decimal. Here's what a place value chart from millions to millionths looks like.

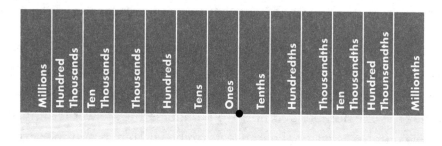

EXAMPLE

▶ What is the value of the digit 8 in 5,173.2485?

▶ Write the number 5,173.2485 in the place value chart.

Thousands	Hundreds	Tens	Ones	Tenths	Hundredths	Thousandths	Ten Thousandths	Hundred Thousandths	Millionths
5	1	7	3 .	2	4	8	5		

▶ Identify the place name that corresponds with the digit 8. The digit 8 appears in the thousandths place. Therefore, 8 times 0.001 equals 0.008.

▶ The 8 in 5,173.2485 is in the thousandths place and represents 0.008.

Rational Numbers

Rational numbers are numbers that can be expressed in the form of a ratio between two numbers, where the denominator is not 0. It's easy to see that any fraction is a rational number since its numerator is one whole number and its denominator is another. Together, the two numbers form a ratio.

When we express the idea that $\frac{3}{5}$ of the students in a class are girls, we are really saying that 3 out of every 5 students in the class are girls. The ratio of girls to total students can be expressed as "3 to 5", 3/5, 0.60, or 60%.

Why are whole numbers such as 3, 943 or 128,478 rational numbers when they are one number and not two numbers? The answer is simple! Any whole number is rational because it can be written as a numerator over a denominator of 1. Thus, 3 equals $\frac{3}{1}$, 943 equals $\frac{943}{1}$, and 128,478 equals $\frac{128,478}{1}$. Although whole numbers are always positive, we generally don't write them with a positive sign (+) before them.

If all positive whole numbers and fractions are rational numbers, what about negative numbers such as -5 or $\dfrac{-1}{2}$? Negative numbers frequently show up in daily life. Think about a temperature of $-5°F$ or a stock price that is down $\dfrac{-1}{2}$ point.

Negative numbers are *always* written with a negative sign before them. (In case you're wondering, 0 is unique in that it is considered neither positive nor negative.) Taken together, all the positive whole numbers and their negative opposites—together with zero—form the set of numbers called **integers**. We can show integers on a number line:

Integer Number Line

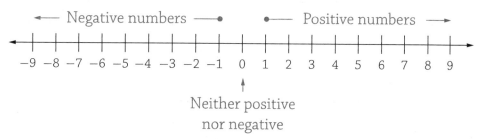

Neither positive
nor negative

Another important way to think about integers is to consider their absolute value. The **absolute value** of a number is its distance from 0 on the number line. When we want to write the absolute value of a number, we place the number between straight lines. Thus, $|4|$ equals 4 and $|-4|$ also equals 4, since both are 4 units from 0. It's important to distinguish between $|-4|$ and $-|4|$. The value of the first number is 4, but that of the second number is -4, since the negative sign is outside the absolute value lines.

▶ Identify each pair of opposite integers on the number line below. Then write the absolute value of each pair of integers.

▶ The pairs of opposite integers on the number line are -5 and 5, -4 and 4, -3 and 3, -2 and 2, and -1 and 1.

▶ The absolute values of each pair of numbers are written as $|5|$, $|4|$, $|3|$, $|2|$, and $|1|$.

Now, we can offer a full, formal definition of rational numbers:

Rational numbers are numbers that can be written in the form of $\frac{a}{b}$, where $b \neq 0$.

All integers and fractions are rational. As for decimals, only those decimals that stop, or terminate, and those that repeat indefinitely are rational. For example, $\frac{1}{4}$ and $\frac{1}{3}$ are both rational, since $\frac{1}{4}$ terminates $\left(\frac{1}{4} = 0.25\right)$ and $\frac{1}{3}$ keeps repeating $\left(\frac{1}{3} = 0.3333333333...\right)$.

Irrational Numbers

Irrational numbers *cannot* be written as the ratio of two integers, and their decimal forms neither terminate nor repeat. At first we might think that irrational numbers occur infrequently. In fact, they are quite commonplace, since the square root of any nonperfect square is irrational. What exactly does this mean? Well, perfect squares include numbers such as 4, 9, and

16 that have both positive and negative square roots that are integers. For example, $\sqrt{4}$ equals 2 and -2, $\sqrt{9}$ equals 3 and -3, and $\sqrt{16}$ equals 4 and -4. The square roots of nonperfect squares such as $\sqrt{2}, \sqrt{3}$, and $\sqrt{5}$ result in nonterminating, nonrepeating decimals.

All rational numbers and all irrational numbers taken together form the set of **real numbers**.

EXAMPLE

▶ Which numbers below are rational numbers and which are irrational numbers?

$\sqrt{11}, \sqrt{16}, \sqrt{20}, \sqrt{25}, \sqrt{30}, \sqrt{36}, \sqrt{39}, \sqrt{49}, \sqrt{64}, \sqrt{75}$

▶ All perfect squares are rational numbers. Therefore, $\sqrt{16}, \sqrt{25}, \sqrt{36},$ $\sqrt{49}$, and $\sqrt{64}$ are rational since their square roots, respectively, are $\pm 4, \pm 5, \pm 6, \pm 7,$ and ± 8.

▶ Since 11, 20, 30, 39, and 75 are nonperfect squares, their square roots are irrational.

▶ $\sqrt{16}, \sqrt{25}, \sqrt{36}, \sqrt{49}$, and $\sqrt{64}$ are rational numbers, and $\sqrt{11}, \sqrt{20}, $ $\sqrt{30}, \sqrt{39}$, and $\sqrt{75}$ are irrational numbers.

 IRL The ancient Greek mathematician Hippasus (one of Pythagoras's students) is said to have discovered the existence of irrational numbers when trying to determine a ratio to express a hypotenuse whose length was $\sqrt{2}$. Instead, Hippasus proved that $\sqrt{2}$ cannot be written as a fraction.

We can compare the value of two real numbers just as we would whole numbers. For example, $\dfrac{2}{5}$ is less than $\dfrac{3}{5}$. Notice that both fractions have the same denominator, so all that has to be done is compare the numerators. Since 2 is less than 3, $\dfrac{2}{5}$ is less than $\dfrac{3}{5}$.

Sometimes, the real numbers we are comparing are expressed in different forms, for example, as 0.35 and $\frac{3}{10}$. To determine which of these numbers is greater, we must represent both in the same form. As we've seen, converting fractions to decimals is easy: All we must do is divide the numerator by the denominator, so $\frac{3}{10}$ equals 0.30. Now, comparison is quick and easy: 0.35 > 0.30.

Number lines are helpful when we want to compare and order the value of several real numbers. For example, suppose we are asked to order the following numbers from least to greatest: -2, 1.25, 3, $\frac{3}{4}$, -3.5, and $\frac{-3}{5}$. Locating points on a number line that represent these numbers makes the ordering easy.

Notice that two of the numbers are written as integers, two as decimals, and two as fractions. Three of the numbers are positive numbers, and three are negative numbers. The first thing to do is to make sure that all the numbers are expressed in the same form. Finding the decimal value of the integers is simple: -2 equals -2.0 and 3 equals 3.0. We already know that to find the decimal value of a fraction: We divide the numerator by the denominator, so $\frac{3}{4}$ equals 0.75 and $\frac{-3}{5}$ equals -0.6.

Last, we must mark the points on the number line that represent the six numbers and then compare their locations. Moving from left to right, the numbers get larger:

From least to greatest, the numbers are ordered as: $-3.5 < -2 < \frac{-3}{5} < \frac{3}{4} < 1.25 < 3$.

Now, let's work on one together.

▶ Write these numbers in order from greatest to least: $\dfrac{-4}{5}$, -2, 1.75, -1.5, $\dfrac{7}{10}$, 3. Use a number line.

▶ Express all numbers so they are in decimal form.

$$\frac{-4}{5} = -0.80$$

$$-2 = -2.0$$

$$\frac{7}{10} = 0.70$$

$$1.75 = 1.75 \qquad \text{(already in decimal form)}$$

$$-1.5 = -1.5 \qquad \text{(already in decimal form)}$$

$$3 = 3.0$$

▶ Mark the numbers on a number line.

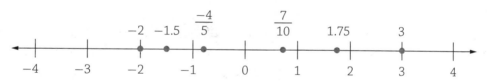

▶ In order from greatest to least: $3 > 1.75 > \dfrac{7}{10} > \dfrac{-4}{5} > -1.5 > -2$.

We must be careful when comparing the values of decimals and fractions that happen to share digits.

▶ Arrange the following numbers in order from least to greatest: $\dfrac{6}{7}$, -3.7, $\dfrac{7}{9}$, -0.57, 3.6, $\dfrac{-5}{7}$. Use a number line.

▶ Find the decimal form of all the numbers. Approximate the numbers to the hundredths place.

$$\frac{6}{7} \approx 0.86$$

$$\frac{7}{9} \approx 0.78$$

$$\frac{-5}{7} \approx -0.71$$

▶ List the numbers from least to greatest beginning at the left and moving right. So, the order of the numbers from least to greatest is:

$$-3.7 < \frac{-5}{7} < -0.57 < \frac{7}{9} < \frac{6}{7} < 3.6$$

Adding Integers

We can use a number line to model the addition of integers. For example, if we want to find the sum of $-2 + -4$, we start at 0 and move 2 units left to represent -2. To add -4, we must move 4 more units left:

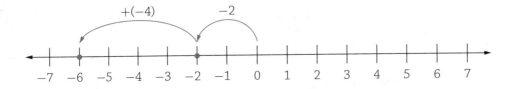

The sum of -2 plus -4, then, equals -6.

Here's an example of how to add a negative integer and a positive integer.

EXAMPLE

▶ Use a number line to find the sum of 6 and −8.

▶ Start at 0 and move 6 units right. Then move 8 units left. Your final position on the number line is −2.

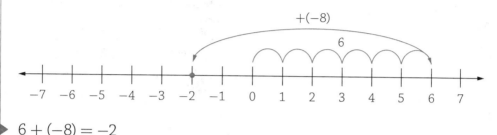

▶ $6 + (-8) = -2$

In the example below, notice where we start and where we end.

EXAMPLE

▶ Use a number line to find the sum of −7 and 7.

▶ Start at 0 and move 7 units left. Then move 7 units right. So, your final position on the number line is 0.

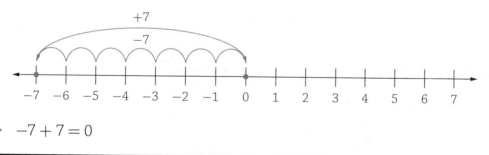

▶ $-7 + 7 = 0$

We can add three or more integers by finding the sum of two integers in order from left to right. What is the sum of 9, −2, 11, and −4?

$$9 + -2 + 11 + -4$$
$$= 7 + 11 + -4$$
$$= 18 + -4$$
$$= 14$$

Another way to solve the problem is by applying the commutative and associative properties of addition. The **commutative property of addition** says that the order in which we add numbers does not change their sum. The **associative property of addition** states that the sum of two or more numbers does not depend on how they are grouped. Let's look again at the same numbers:

$$9 + -2 + 11 + -4 = 9 + 11 + -2 + -4 \quad \text{Use the commutative property.}$$
$$= (9 + 11) + (-2 + -4) \quad \text{Use the associative property.}$$
$$= 20 + -6$$
$$= 14$$

Let's consider an example that involves the addition of more than two integers.

EXAMPLE

▶ A football team gained 5 yards, lost 10 yards, gained 2 yards, lost 8 yards, and then lost 2 yards. What was the football team's net loss or net gain?

▶ Write the problem using integers based on the sequence of gains and losses described in the problem.

$$5 + -10 + 2 + -8 + -2 = ?$$

▶ Find the sum by adding pairs of numbers from left to right.

$$5 + -10 + 2 + -8 + -2$$
$$= -5 + 2 + -8 + -2$$
$$= -3 + -8 + -2$$
$$= -11 + -2$$
$$= -13$$

▶ The football team had a net loss of 13 yards.

We can use what we know about absolute value to add two or more integers. To add two integers with the same sign, add their absolute values and use the sign of the numbers.

EXAMPLE

▶ Find the sum of −3 and −11.

▶ Write the absolute value of each number and then add them.

$$|-3| + |-11| = 3 + 11 = 14$$

▶ Since both of the original numbers are negative, place a negative sign in front of the sum.

▶ Therefore, the sum of −3 and −11 equals −14.

We can summarize the rules for adding integers:

Same Signs	Add the absolute values and use the common sign of the addends.	$6 + 3 = 9$ $-3 + -4 = -7$
Different Signs	Subtract the lesser absolute value from the greater absolute value. Use the sign of the number with greater absolute value.	$11 + -5 = 6$ $-9 + 4 = -5$
Sum of Opposites	The sum of an integer and its opposite is 0.	$-7 + 7 = 0$ $3 + -3 = 0$

Subtracting Integers

We can also use a number line to model the subtraction of integers. Remember that when we add a positive integer, we must move right. To subtract a positive integer, we move to the left. For example, if we want to find the difference between 8 and 3, we can start at 0 and move 8 units right to represent $+8$. To subtract 3, we must move 3 units left.

Subtract an integer by adding its opposite.	$8 - 3 = 8 + {-3} = 5$

Let's use a number line to show how to subtract a negative integer from a positive integer.

▶ Use a number line to find the difference between 5 and -2.

▶ Start at 0 and move 5 units right. Since $-(-2)$ equals $+2$, move 2 units right. Note that your final position on the number line is 7.

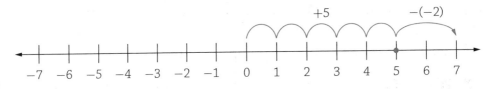

▶ $5 - (-2) = 7$

The example below shows how to subtract a negative integer from a negative integer.

▶ What is the difference between -5 and -4?

▶ Find the absolute value of the numbers.

$$|-5| = 5 \text{ and } |-4| = 4$$

▶ Subtract the lesser number from the greater number: $5 - 4 = 1$.

▶ Add the sign of the numbers to the difference: -1.

▶ The difference between -5 and -4 is -1.

Here's a problem that involves the subtraction of two negative integers.

▶ The top of a coral reef is 410 feet below the surface of the water. The base of the coral reef is 625 feet below the water's surface. What is the difference between the top of coral reef and its base?

▶ Write an equation to reflect the facts presented in the problem.

$$-410 - (-625) = ?$$

▶ Subtract the numbers. Notice the change in sign when you subtract a negative number.

$$-410 + 625 = 215$$

▶ The difference from the top of the coral reef to its base is 215 feet.

Multiplying Integers

Recall that addition and multiplication are related operations, since multiplication can be thought of as repeated addition. Our understanding of the addition of integers can help us understand the rules for multiplying integers:

Same Signs	The product of two integers with the same sign is positive.	$2 \times 4 = 8$ $-2 \times -4 = 8$
Different Signs	The product of two integers with different signs is negative.	$-2 \times 4 = -8$ $2 \times -4 = -8$
Zero as a Factor	The product of an integer and 0 is 0.	$4 \times 0 = 0$ $-4 \times 0 = 0$

Let's consider some problems that involve the multiplication of integers.

EXAMPLE

▶ Find the product of each problem. Name the rule you used to solve each problem.

a. -5×9

b. -2×-12

c. 3×-4

d. -8×0

▶ a. -45 The product of two integers with different signs is negative.

b. 24 The product of two integers with the same sign is positive.

c. -12 The product of two integers with different signs is negative.

d. 0 The product of an integer and 0 is 0.

Here's a word problem based on the multiplication of integers.

▶ Scientists studying the hiberation of bears found that, on average, a bear loses about six pounds per month while asleep. If the period of hiberation was 5 months, how many pounds did the bear's weight drop?

▶ Use integers to write the problem as an equation.

$$-6 \times 5 = ?$$

▶ Multiply the numbers and determine the sign. Since the two factors have opposite signs, the product is negative.

$$-6 \times 5 = -30$$

▶ The bear's weight dropped 30 pounds.

Dividing Integers

Recall that multiplication and division are inverse operations. Therefore, we can use what we know about multiplication to help us understand the rules for dividing integers:

Same Signs	The quotient of two integers with the same sign is positive.	$12 \div 4 = 3$ $-12 \div -4 = 3$
Different Signs	The quotient of two integers with different signs is negative.	$12 \div -4 = -3$ $-12 \div 4 = -3$
Zero as a Factor	The quotient of 0 and any nonzero integer is 0.	$0 \div 12 = 0$ $0 \div -12 = 0$

Let's practice the division of integers with the following problems.

BTW

Always remember that we can never divide a number by 0!

EXAMPLE

Find the quotient of each problem. Name the rule you used to solve the problem.

a. $56 \div -7$

b. $-22 \div 2$

c. $(-48) \div (-4)$

d. $0 \div -8$

a. $56 \div -7 = -8$ The quotient of two integers with different signs is negative.

b. $-22 \div 2 = -11$ The quotient of two integers with different signs is negative.

c. $-48 \div -4 = 12$ The quotient of two integers with the same sign is positive.

d. $0 \div -8 = 0$ The quotient of 0 divided by any integer is 0.

EXERCISES

EXERCISE 1–1

Identify the value of:

1. 3 in the number 1,321.625

2. 5 in the number 51,213.479

3. 4 in the number 723.645

4. 9 in the number 201.439

EXERCISE 1–2

Identify the absolute value of each integer.

1. −3

2. 2

3. 4

4. −5

EXERCISE 1–3

Identify each number as rational or irrational.

1. $\dfrac{4}{9}$

2. π

3. −8

4. $\sqrt{12}$

EXERCISE 1–4

Compare the indicated numbers using the symbols >, <, or =.

1. -2 and -5

2. 3 and -4

3. 0.2 and $\dfrac{1}{5}$

4. $\dfrac{6}{9}$ and 0.50

EXERCISE 1–5

Order the numbers as indicated.

1. from least to greatest: $-4, \dfrac{2}{3}, -2, |3.5|, \dfrac{-2}{5}, 0.25$

2. from greatest to least: $0.35, \dfrac{-5}{8}, 5.7, 4, -5.2, -5$

3. from least to greatest: $\sqrt{14}, 3, -3.5, -0.75, 3.2, \dfrac{-4}{7}$

4. from greatest to least: $-1.7, \dfrac{-72}{-12}, \sqrt{35}, 5.8, -0.45, \dfrac{-9}{10}$

EXERCISE 1–6

Find the sum of the integers.

1. $2 + -9$

2. $-15 + -11$

3. $13 + -7 + 3$

4. $-9 + 5 + -4 + 2$

EXERCISE 1–7

Find the difference between the integers.

1. $3 - (-9)$

2. $-12 - (-10)$

3. $-5 - (-8) - (-3)$

4. $-7 - (-2) - (-5)$

EXERCISE 1–8

Find the product of the integers.

1. -4×-6

2. 8×-9

3. $-13 \times 0 \times -5$

4. $-7 \times -4 \times -1$

EXERCISE 1–9

Find the quotient of the integers.

1. $56 \div -7$

2. $\dfrac{-72}{-12}$

3. $-65 \div 5$

4. $-81 \div -9$

EXERCISE 1–10

Solve each word problem using what you know about integers.

1. Melinda's watch loses 4 minutes per day. How many minutes will her watch lose in two weeks?

2. Jasper bought 200 shares of a stock. A month later he had lost $800 on the investment. In dollars, what was the loss per share over the month?

3. On its first day, a new stock was issued at $10 per share. By noon, the stock had dropped $8 from its original price. By 2 P.M., the price had increased by $6. At the stock market's 3 P.M. closing, the stock's price was $5 less than its price at 2 P.M. What was the stock's closing price?

4. The temperature was 15°F at 6 A.M. on Monday. At 6 A.M. on Tuesday the temperature was −11°F. How many degrees Farenheit did the temperature drop from Monday to Tuesday?

Fractions and Decimals

ecimals and fractions are all around us in daily life. Every time we deal with money, we are also dealing with decimals. $1.10 means one dollar and one-tenth of a dollar. A quarter is worth 25 cents, or $0.25. Likewise, when we go shopping, we may ask for $\frac{1}{2}$ pound of sliced turkey, although the label on the package will likely say 0.50 pound. The next time you are out and about, keep track of times you encounter decimals and fractions. You're likely to be surprised by how frequently you find them.

Fractions

A **fraction** is a rational number that expresses a ratio of the number of equal parts we have to the number of equal parts that represent a whole. The top number, or **numerator**, tells us how many parts we have. The bottom number, or **denominator**, tells us how many parts make up the whole.

Suppose a circle is divided into five equal parts of which two parts are shaded:

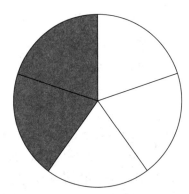

Based on the diagram, we can say that $\frac{2}{5}$ of the circle is shaded and $\frac{3}{5}$ is unshaded. Notice that $\frac{2}{5}$ plus $\frac{3}{5}$ equal the whole, or 1:

$$\frac{2}{5} + \frac{3}{5} = \frac{5}{5} = 1$$

Every fraction can be expressed as a decimal. To find the decimal value of a fraction, all we need to do is divide the numerator by the denominator. Based on the circle shown on the preceding page,

$$\frac{2}{5} = 0.4 \text{ and } \frac{3}{5} = 0.6$$

If we add 0.4 and 0.6, the sum is 1.0. So, like the fractions that represent the shaded and unshaded parts of the circle, the decimal representations of these parts also have a sum of 1.

Notice that the decimal representations of the fractions $\frac{2}{5}$ and $\frac{3}{5}$ terminate, or come to an end. That's not the case with the decimal representations of all fractions. Some, for example, $\frac{1}{7}$ and $\frac{1}{11}$, repeat indefinitely:

$$\frac{1}{7} = 0.\overline{142857} \text{ and } \frac{1}{11} = 0.\overline{09}$$

In a decimal number, a bar above consecutive digits means that the pattern of digits under the bar repeats without end. Thus, $0.\overline{09}$ means 0.09090909....

▶ Write the shaded number of parts in the diagram as a fraction and as a decimal.

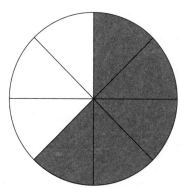

▶ The diagram shows $\dfrac{5}{8}$ of the diagram shaded, and 5 divided by 8 equals 0.625.

▶ Therefore, $\dfrac{5}{8}$ equals 0.625.

Here's an example that represents a fraction as part of a set of objects.

EXAMPLE

▶ Write the shaded number of squares in the diagram as a fraction and as a decimal.

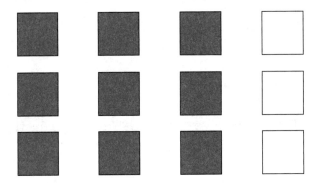

▶ The diagram shows $\dfrac{9}{12}$ of the squares shaded, and 9 divided by 12 equals 0.75.

▶ So, $\dfrac{9}{12}$ equals 0.75.

Equivalent Fractions

Fractions that represent the same part-to-whole relationship are called **equivalent fractions**. For example, $\dfrac{3}{4}$ and $\dfrac{15}{20}$ are equivalent fractions. We can prove this by reducing $\dfrac{15}{20}$ to its simplest form by dividing both the numerator and denominator by 5.

The diagram in the previous example shows $\dfrac{9}{12}$ of the parts shaded. We can create an equivalent fraction by dividing both the numerator and denominator by 3:

$$\frac{9 \div 3}{12 \div 3} = \frac{3}{4}$$

We can create equivalent fractions by using multiplication. If we wanted to create two fractions that are equivalent to $\dfrac{5}{9}$, we could multiply the numerator and denominator by the same number. For example,

$$\frac{5 \times 2}{9 \times 2} = \frac{10}{18} \text{ and } \frac{5 \times 3}{9 \times 3} = \frac{15}{27}$$

So, $\dfrac{5}{9}$ and $\dfrac{10}{18}$ are equivalent fractions, and $\dfrac{5}{9}$ and $\dfrac{15}{27}$ are also equivalent fractions. Since both $\dfrac{10}{18}$ and $\dfrac{15}{27}$ are equal, we can also say that *they* are equivalent fractions. In fact, using this method, we can create an infinite number of fractions equivalent to $\dfrac{5}{9}$.

EXAMPLE

▶ What are two fractions that are equivalent to $\dfrac{4}{5}$?

▶ The first thing we have to do is choose a number to multiply both the numerator and the denominator by. If we choose the number 2, we can determine an equivalent fraction as shown in the following example.

$$\frac{4}{5} = \frac{4 \times 2}{5 \times 2} = \frac{8}{10}$$

▶ If we choose a different number, say 3, then we can determine a different equivalent fraction to $\dfrac{4}{5}$.

$$\frac{4}{5} = \frac{4 \times 3}{5 \times 3} = \frac{12}{15}$$

▶ Both $\dfrac{8}{10}$ and $\dfrac{12}{15}$ are equivalent fractions to $\dfrac{4}{5}$.

Sometimes, we need to find a missing numerator or denominator to find equivalent fractions. This is actually the same situation we face when we find the missing term in a proportion. To do this, we must determine the relationship that exists between the two numerators or two denominators that appear in the problem.

EXAMPLE

▶ $\dfrac{5}{7} = \dfrac{20}{?}$

▶ Think: Examine the numerators.

What number times 5 equals 20?
$5 \times 4 = 20$
So, what does 7 times 4 equal?
$\rightarrow 7 \times 4 = 28$

▶ Therefore, $\dfrac{5}{7}$ equals $\dfrac{20}{28}$.

The next example shows how to find a missing numerator in equivalent fractions.

EXAMPLE

▶ Find the fraction that is equivalent to $\dfrac{40}{56}$.

▶ Find the relationship between the two known denominators.

$$\frac{40}{56} = \frac{?}{7}$$

▶ Think: Examine the denominators.

What number do you have to divide 56 by to get the quotient 7?
$56 \div 8 = 7$
Apply this fact to the numerators: $40 \div 8 = 5$.
Apply the relationship to find the unknown numerator: $40 \div 8 = 5$.

▶ Therefore: $\dfrac{40 \div 8}{56 \div 8} = \dfrac{5}{7}$.

▶ So, $\dfrac{40}{56}$ and $\dfrac{5}{7}$ are equivalent fractions.

A fraction is in its **simplest form** if its numerator and denominator have only 1 as a common factor. In other words, no number other than 1 can divide both the numerator and denominator. To determine if this is the case, we need to know the **greatest common factor** (GCF) of the numerator and denominator. Recall that the GFC of two or more whole numbers is the largest positive integer that evenly divides all the numbers with zero remainders. To find the GFC of 18 and 42, we would make a list of the factors of both numbers:

Factors of 18: 1, 2, 3, **6**, 9, 18
Factors of 42: 1, 2, 3, **6**, 7, 14, 21

EXAMPLE

▶ Write the fraction $\dfrac{21}{36}$ in its simplest form.

▶ Write the original fraction: $\dfrac{21}{36}$.

▶ Find the GCF of the numerator and denominator.

Factors of 21: 1, **3**, 7, 21
Factors of 36: 1, 2, **3**, 4, 9, 12, 18, 36

▶ The GCF is 3, so we divide the numerator and denominator by 3.

$$\frac{21 \div 3}{36 \div 3} = \frac{7}{12}$$

▶ $\frac{21}{36}$ in its simplest form is $\frac{7}{12}$.

Finding the simplest form of a fraction helps us understand the relationship between the part and the whole more easily. Let's look at a word problem that shows why.

▶ In a survey of 75 teenagers, 35 think school A will win the state basketball championship over school B. Write the results of the survey as a fraction in its simplest form.

▶ Write the original fraction: $\frac{35}{75}$.

▶ Divide the numerator and denominator by the GCF of both numbers: $\frac{35}{75}$. The GCF of 35 and 75 is 5.

$$\frac{35}{75} = \frac{35 \div 5}{75 \div 5} = \frac{7}{15}$$

▶ In simplest form, the fraction of students who think school A will win the state basketball championship over school B is $\frac{7}{15}$. Therefore, slightly less than half the students think school A will win the state championship.

Comparing and ordering fractions with different denominators requires care. The first thing we must do is express all the fractions with the same denominator. To do, we must find the least common denominator of the fractions we are comparing. The **least common denominator** (LCD) is the lowest number that we can use to create a set of fractions that all have the same denominator.

EXAMPLE

▶ Which is greater: $\dfrac{2}{3}$ or $\dfrac{3}{5}$?

▶ Since $\dfrac{2}{3}$ and $\dfrac{3}{5}$ do not have the same denominator, they cannot be readily compared. To make the comparison we must find the LCD.

▶ To find the LCD of the fractions, we must find the least common multiple (LCM) of the denominators. The least common multiple (LCM) is the smallest positive integer that is divisible by the two integers. To find the LCM of 3 and 5, list the multiples of 3 and 5.

> Multiples of 3: 3, 6, 9, 12, **15**, 18, 21, ...
> Multiples of 5: 5, 10, **15**, 20, 25, 30, ...

▶ The *LCD* of the fractions is 15.

▶ Write equivalent fractions using the LCD.

$$\frac{2}{3} = \frac{2 \times 5}{3 \times 5} = \frac{10}{15}$$

$$\frac{3}{5} = \frac{3 \times 3}{5 \times 3} = \frac{9}{15}$$

▶ Compare the two fractions. $\dfrac{10}{15}$ is greater than $\dfrac{9}{15}$.

▶ Simplify the fractions. $\dfrac{2}{3}$ is greater than $\dfrac{3}{5}$.

We can use the same procedure to order multiple numbers from least to greatest or vice versa.

▶ Order the fractions $\dfrac{2}{3}$, $\dfrac{3}{4}$, and $\dfrac{7}{12}$ from least to greatest.

▶ Find the LCD of the fractions. The LCD of 3, 4, and 12 is 12.

▶ Write equivalent fractions so they all have the same LCD.

$$\frac{2}{3} = \frac{2 \times 4}{3 \times 4} = \frac{8}{12}$$

$$\frac{3}{4} = \frac{3 \times 3}{4 \times 3} = \frac{9}{12}$$

$$\frac{7}{12}$$

▶ In order from least to greatest, the fractions are: $\dfrac{7}{12} < \dfrac{2}{3} < \dfrac{3}{4}$.

Adding and Subtracting Fractions

To add and subtract fractions with like denominators is as easy as one, two, three. First, add or subtract the numerators. Second, keep the denominator. Third, write the fraction in simplest form.

▶ What is the sum of $\dfrac{11}{25}$ and $\dfrac{9}{25}$? Write the sum in its simplest form.

▶ Write the original problem.

$$\frac{11}{25} + \frac{9}{25} = ?$$

▶ Add the numerators. Keep the denominator the same.

$$\frac{11}{25} + \frac{9}{25} = \frac{20}{25}$$

▶ Reduce the fraction to its simplest form.

$$\frac{20}{25} = \frac{20 \div 5}{25 \div 5} = \frac{4}{5}$$

▶ Reduced to its simplest form, the sum of $\frac{11}{25}$ and $\frac{9}{25}$ is $\frac{4}{5}$.

The same steps can be used when subtracting two fractions with like denominators.

EXAMPLE

▶ What is the difference between $\frac{13}{15}$ and $\frac{8}{15}$? Write the difference in its simplest form.

▶ Write the original problem.

$$\frac{13}{15} - \frac{8}{15} = ?$$

▶ Subtract the numerators. Keep the denominator the same.

$$\frac{13}{15} - \frac{8}{15} = \frac{5}{15}$$

▶ Simplify the answer to lowest terms.

$$\frac{13}{15} - \frac{8}{15} = \frac{5}{15} = \frac{5 \div 5}{15 \div 5} = \frac{1}{3}$$

▶ Reduced to its simplest form, the difference between $\frac{13}{15}$ and $\frac{8}{15}$ is $\frac{1}{3}$.

When we add and subtract fractions with unlike denominators, we first convert them to fractions with the same denominators. Then, we perform the addition or subtraction. Finally, we write the fraction in its simplest form. Let's look at a problem that involves finding the sum of two fractions with unlike denominators.

▶ What is the sum of $\dfrac{3}{5}$ and $\dfrac{1}{4}$?

▶ Write the original problem.

$$\frac{3}{5} + \frac{1}{4} = ?$$

▶ Write the fractions with a common denominator.

$$\frac{3 \times 4}{5 \times 4} = \frac{12}{20}$$

$$\frac{1 \times 5}{4 \times 5} = \frac{5}{20}$$

▶ Add the fractions.

$$\frac{12}{20} + \frac{5}{20} = \frac{17}{20}$$

▶ $\dfrac{17}{20}$ is already in its simplest form.

▶ Reduced to its simplest form, the sum of $\dfrac{3}{5}$ and $\dfrac{1}{4}$ is $\dfrac{17}{20}$.

The example that follows shows how to subtract fractions that have different denominators. The procedure is similar to the one used when we add fractions with unlike denominators. We first write equivalent fractions with the same denominators, subtract them, and, if necessary, reduce the difference to its simplest form.

▶ What is the difference between $\dfrac{1}{2}$ and $\dfrac{3}{10}$? Write the difference in its simplest form.

▶ Write the original problem.

$$\dfrac{1}{2} - \dfrac{3}{10} = ?$$

▶ Write the fractions with a common denominator.

$$\dfrac{1 \times 5}{2 \times 5} = \dfrac{5}{10}$$

▶ Subtract the fractions.

$$\dfrac{5}{10} - \dfrac{3}{10} = \dfrac{2}{10}$$

▶ Simplify the answer to lowest terms.

$$\dfrac{2}{10} = \dfrac{2 \div 2}{10 \div 2} = \dfrac{1}{5}$$

▶ Reduced to its simplest form, the difference between $\dfrac{1}{2}$ and $\dfrac{3}{10}$ is $\dfrac{1}{5}$.

Adding and Subtracting Mixed Numbers

Some problems involve adding or subtracting mixed numbers. A **mixed number** consists of a whole number and a fractional part. In order to perform addition and subtraction, we must change the mixed number into an improper fraction. An **improper fraction** is any fraction in which the numerator is greater than the denominator.

▶ This past weekend Francine jogged $2\frac{3}{4}$ miles on Saturday and $1\frac{4}{5}$ miles on Sunday. What is the total distance Francine jogged this past weekend?

▶ Write the original problem.

$$2\frac{3}{4} + 1\frac{4}{5} = ?$$

▶ Write each addend as an improper fraction.

$$2\frac{3}{4} = \frac{11}{4}$$

$$1\frac{4}{5} = \frac{9}{5}$$

▶ Find the LCD of the fractions $\frac{11}{4}$ and $\frac{9}{5}$.

$$\frac{11 \times 5}{4 \times 5} = \frac{55}{20}$$

$$\frac{9 \times 4}{5 \times 4} = \frac{36}{20}$$

▶ Add the renamed fractions.

$$\frac{55}{20} + \frac{36}{20} = \frac{91}{20}$$

▶ Simplify the sum and write it as a mixed number.

$$\frac{91}{20} = 4\frac{11}{20}$$

▶ Francine jogged $4\frac{11}{20}$ miles this past weekend.

Let's take a look at another method for adding or subtracting mixed numbers with different denominators in the fractions. The process involves changing both numbers to improper fractions, subtracting, and then simplifying the difference.

▶ What is the difference between $5\dfrac{2}{5}$ and $1\dfrac{1}{3}$?

▶ Write the original problem.

$$5\frac{2}{5} - 1\frac{1}{3} = ?$$

▶ Rename the mixed numbers as improper fractions and rewrite the problem.

$$5\frac{2}{5} = \frac{27}{5}$$

$$1\frac{1}{3} = \frac{4}{3}$$

$$\downarrow$$

$$\frac{27}{5} - \frac{4}{3} = ?$$

▶ Find the LCD and rename the fractions.

$$\frac{27 \times 3}{5 \times 3} = \frac{81}{15}$$

$$\frac{4 \times 5}{3 \times 5} = \frac{20}{15}$$

▶ Subtract the improper fractions.

$$\frac{81}{15} - \frac{20}{15} = \frac{61}{15}$$

▶ Simplify the difference.

$$\frac{61}{15} = 4\frac{1}{15}$$

▶ The difference between $5\frac{2}{5}$ and $1\frac{1}{3}$ is $4\frac{1}{15}$.

BTW

As we've seen, converting a mixed number into an improper fraction makes it easier to solve problems. However, an improper fraction in an answer should always be simplified into a mixed number.

Multiplying Fractions and Mixed Numbers

The steps for multiplying fractions and mixed numbers are easy to follow:

1. Write all the whole numbers and mixed numbers as fractions.
2. Multiply the numerators of the fractions.
3. Multiply the denominators of the fractions.
4. Write the product in its simplest form.

EXAMPLE

▶ What is the product of $\frac{3}{5}$ and $\frac{2}{3}$? Write the product in its simplest form.

▶ Write the original problem.

$$\frac{3}{5} \times \frac{2}{3} = ?$$

▶ Multiply the numerators and then the denominators.

$$\frac{3}{5} \times \frac{2}{3} = \frac{6}{15}$$

▶ Reduce the product to its simplest form.

$$\frac{6 \div 3}{15 \div 3} = \frac{2}{5}$$

▶ The product of $\frac{3}{5}$ and $\frac{2}{3}$ in simplest form is $\frac{2}{5}$.

Remember that we can write any whole number as a fraction by writing the number over a denominator of 1—for example, 5 equals $\frac{5}{1}$ and 12 equals $\frac{12}{1}$. This fact comes in handy when we are multiplying a fraction and a whole number.

BTW

Often, we can make the multiplication of fractions easier by reducing fractions to their simplest form before multiplying.

Since $\frac{5}{15}$ equals $\frac{1}{3}$

and $\frac{3}{6}$ equals $\frac{1}{2}$, we can mentally compute:

$$\frac{1}{3} \times \frac{1}{2} = \frac{1}{6}.$$

That's a lot easier than multiplying:

$$\frac{5}{15} \times \frac{3}{6} = \frac{15}{90} = \frac{1}{6}.$$

EXAMPLE

▶ There are 30 students in the drama club, and $\frac{3}{5}$ of the members are girls. How many girls are in the drama club?

▶ Write the original problem. Recall that you can express any integer as a fraction by writing it as a numerator over the denominator 1.

$$\frac{3}{5} \times \frac{30}{1} = ?$$

▶ Multiply the numerators and then the denominators.

$$\frac{3}{5} \times \frac{30}{1} = \frac{90}{5}$$

▶ Reduce the product to lowest terms.

$$\frac{90 \div 5}{5 \div 5} = \frac{18}{1} = 18$$

▶ The product of $\frac{3}{5}$ and 30 is 18.

When a problem asks us to multiply **mixed numbers**, we must first rewrite the numbers as improper fractions and then follow the same rules for multiplying fractions. Recall that an improper fraction is a fraction in which the numerator is greater than the denominator.

EXAMPLE

▶ What is the product of $2\frac{1}{3}$ and $4\frac{3}{4}$?

▶ Change the mixed numbers into improper fractions.

$$2\frac{1}{3} = \frac{7}{3}$$

$$4\frac{3}{4} = \frac{19}{4}$$

▶ Multiply the numerators and then the denominators.

$$\frac{7 \times 19}{3 \times 4} = \frac{133}{12}$$

▶ Reduce the product to simplest form.

$$\frac{133}{12} = 11\frac{1}{12}$$

▶ The product of $2\frac{1}{3}$ and $4\frac{3}{4}$, reduced to its simplest form, is $11\frac{1}{12}$.

Dividing Fractions and Mixed Numbers

To divide fractions and mixed numbers, we use the same concepts that we learned for multiplying *plus* one other idea—the reciprocal. Recall that two numbers are **reciprocals** when their product equals 1. For example, $\frac{3}{5}$ and $\frac{5}{3}$ are reciprocals because: $\frac{3}{5} \times \frac{5}{3} = \frac{15}{15} = 1$. To form the reciprocal of any fraction, all we have to do is switch the numerator and denominator. Thus, the reciprocal of $\frac{3}{7}$ is $\frac{7}{3}$ and their product is $\frac{21}{21}$, or 1.

EXAMPLE

▶ What is the quotient of $\frac{2}{3}$ and $\frac{1}{6}$?

▶ Write the original problem.

$$\frac{2}{3} \div \frac{1}{6} = ?$$

▶ Rewrite as a multiplication problem, using the reciprocal of the divisor.

$$\frac{2}{3} \times \frac{6}{1} = ?$$

▶ Multiply the numerators and then the denominators.

$$\frac{2}{3} \times \frac{6}{1} = \frac{12}{3}$$

▶ Reduce the quotient to lowest terms.

$$\frac{12 \div 3}{3 \div 3} = \frac{4}{1} = 4$$

▶ The quotient of $\frac{2}{3}$ and $\frac{1}{6}$, reduced to lowest terms, is 4.

To divide a fraction by a whole number, rewrite the whole number as the numerator and write it over the denominator 1. Dividing mixed numbers can be a little more complicated.

▶ What is the quotient of $\frac{4}{5}$ and 6?

▶ Write the original problem, using 1 as the denominator of the whole number.

$$\frac{4}{5} \div \frac{6}{1} = ?$$

▶ Rewrite as a multiplication problem, using the reciprocal of the divisor.

$$\frac{4}{5} \times \frac{1}{6} = ?$$

▶ Multiply the numerators and then the denominators.

$$\frac{4}{5} \times \frac{1}{6} = \frac{4}{30}$$

▶ Reduce the quotient to lowest terms.

$$\frac{4 \div 2}{30 \div 2} = \frac{2}{15}$$

▶ The quotient of $\frac{4}{5}$ and 6, reduced to lowest terms, is $\frac{2}{15}$.

Dividing mixed numbers is not just a "math" problem. Sometimes the need to divide mixed numbers has practical value, as the example that follows shows.

EXAMPLE

▶ Bret is baking cookies for the school bake sale. He has $11\frac{1}{4}$ cups of sugar. Each batch of cookies uses $2\frac{1}{4}$ cups of sugar. How many batches of cookies can Bret bake?

▶ Write the original problem.

$$11\frac{1}{4} \div 2\frac{1}{4} = ?$$

▶ Rewrite the mixed numbers as improper fractions.

$$11\frac{1}{4} = \frac{45}{4}$$

$$2\frac{1}{4} = \frac{9}{4}$$

▶ Rewrite the problem as a multiplication problem. Use the reciprocal of the second fraction. Then, multiply the numerators and the denominators.

$$\frac{45}{4} \div \frac{9}{4} = \frac{45}{4} \times \frac{4}{9} = \frac{180}{36}$$

▶ Reduce the quotient to lowest terms.

$$\frac{180 \div 36}{36 \div 36} = \frac{5}{1} = 5$$

▶ Bret can bake 5 batches of cookies.

Sometimes, it's a good idea to estimate the quotient when dividing fractions and mixed numbers. We can do this in the same way we estimate products when multiplying.

▶ What is a reasonable estimate of the quotient of $\frac{11}{12}$ and $\frac{1}{3}$?

▶ Rewrite as a multiplication problem, using the reciprocal of the divisor.

$$\frac{11}{12} \times \frac{3}{1} = ?$$

▶ Round to find approximate whole numbers.

$$\frac{11}{12} \approx 1$$

$$3 = 3$$

▶ Estimate: There are approximately three instances of $\frac{1}{3}$ in $\frac{11}{12}$.

▶ A reasonable estimate of the quotient of $\frac{11}{12}$ and $\frac{1}{3}$ is 3.

Sometimes, a real-world situation may require rounding our exact calculation to arrive at the correct answer.

EXAMPLE

▶ Michelle is making scarves as holiday gifts for five family members. She has $13\frac{1}{2}$ yards of fabric. Each scarf takes about $2\frac{2}{3}$ of a yard of fabric. Does Michelle have enough fabric to make the five scarves?

▶ Write the facts as a division problem.

$$13\frac{1}{2} \div 2\frac{2}{3} = ?$$

▶ Rewrite the mixed numbers as improper fractions.

$$13\frac{1}{2} = \frac{27}{2}$$

$$2\frac{2}{3} = \frac{8}{3}$$

▶ Rewrite the problem as a multiplication problem using the improper fractions. Use the reciprocal of the second fraction. Then, multiply the numerators and the denominators.

$$\frac{27}{2} \times \frac{3}{8} = \frac{81}{16}$$

▶ Reduce the quotient to lowest terms.

$$\frac{81}{6} = 5\frac{1}{16}$$

▶ The quotient is a little more than 5, so Michelle has enough fabric to make 5 scarves.

Decimals

A **decimal** is a number that is written using the base-10 number system. Each place in a decimal is ten times the value of the place to its right. The word *decimal* actually comes from the Latin *deci-*, meaning "ten." A decimal number contains a **decimal point**. The numbers to the left of the decimal point are whole numbers, whereas the numbers to the right of the decimal point are fractional parts. Whole numbers, such as 4 or 121, can be written as the decimals 4.0 or 121.0. And numbers less than 1, such as $\frac{1}{4}$ or $\frac{2}{5}$, can be written as 0.25 and 0.4, respectively.

Familiarizing ourselves with the names of the decimal places is important to understanding and working successfully with decimals. For example, consider the number 9,417.325. Notice that the place names to the right of the decimal point have names similar to the whole numbers at the left but end in the suffix –*th*.

Hundred Thousands	Ten Thousands	Thousands	Hundreds	Tens	Ones	Tenths	Hundredths	Thousandths	Ten Thousandths
		9	4	1	7 .	3	2	5	

We read the number above as "nine thousand, four hundred seventeen and three hundred twenty-five thousandths." 9,417.325 means 9,417 and 325 thousandths:

- The number 9 is in the thousands place, so it represents 9,000.

- The number 4 is hundreds place, and it represents 400.

- The number 1 is in the tens place and represents 10.

- The number 7 is in the ones place, so it stands for 7.

- The number 3 is to the right of the decimal point in the tenths place, and it represents 0.3.

- The 2 is in the hundredths place and represents 0.02.

- Likewise, the number 5 is in the thousandths place and represents 0.005.

In addition to the **decimal form** of a number, decimals also have a word form. As we saw, 9,417.325 is expressed in **word form** as "nine thousand, four hundred seventeen, and three hundred twenty-five thousandths." Another example of the word form of a decimal is "eighty-five hundredths," which stands for 0.85. Note that a 0 is placed before the decimal point when the number is less than 1.

EXAMPLE

▶ What is the decimal form of "two hundred fifty-three and forty-five thousandths"?

▶ Identify the value of the whole number part of the decimal. "Two hundred fifty-three" is written as 253.

▶ Then, identify how many places to the right of the decimal point represents the place value of the fractional part. "Forty-five thousandths" is three places to the right of the decimal point and is written as .045.

▶ So, the correct decimal form of two hundred fifty-three and forty-five hundredths is 253.045.

Let's try reading a decimal aloud together.

▶ How do we read the decimal 734.06?

▶ Write the whole number part in words. 734 is represented by the words "seven hundred thirty-four."

▶ Write the fractional part in words. The decimal .06 is represented by the words "six hundredths."

▶ The word name for 734.06 is "seven hundred thirty-four and six hundredths."

Another way to represent decimals is in **expanded form**. When we write a number in expanded form, we represent each place value in the form of an addition expression. Here's how we would represent 2,436.65 in expanded form:

$$2 \times 1000 + 4 \times 100 + 3 \times 10 + 6 \times 1 + 6 \times 0.1 + 5 \times 0.01$$

Let's write a decimal in expanded together.

▶ How do you write 739.245 in expanded form?

▶ Identify the value of each digit moving from left to right based on its place. Write the value of the digit using place value.

The 7 → 7×100

The 3 → 3×10

The 9 → 9×1

The 2 → $2 \times \dfrac{1}{10}$

The 4 → $4 \times \dfrac{1}{100}$

The 5 → $5 \times \dfrac{1}{1000}$

> 739.245 written in expanded form is:
>
> $$7 \times 100 + 3 \times 10 + 9 \times 1 + 2 \times \frac{1}{10} + 4 \times \frac{1}{100} + 5 \times \frac{1}{1000}$$

Comparing and Ordering Decimals

Sometimes, we want to compare the value of two or more decimals. For example, suppose we want to determine which decimal is greater, 29.764 or 29.798. To compare these two decimals, we should first write them in a column, taking care to line up the decimal points:

29.764
29.798

Then, compare each digit moving from left to right. Notice that the digits in the tens place are the same: 2. Continue to the next place. The digits in the ones place are both 9s. Move to the first place after the decimal point. Here, too, both digits in the tenths place are the same: 7. Next, continue to the hundredths place. In this case, the digits are different.

29.7**6**4
29.7**9**8

Since 9 is greater than 6, 29.798 is greater than 29.764.

EXAMPLE

▶ Complete the statement that follows with <, >, or =.

0.509 _?_ 0.51

▶ To determine which number is greater, write the decimals in a column, lining up the decimal points.

0.509
0.51

▶ If necessary, write zeros to the right of the decimal points so the decimals all have the same number of places.

0.509
0.510

▶ Compare place values from left to right.

0.5**09**
0.5**10**

▶ Think: 0 is less than 1.

▶ 0.509 < 0.51

We can use the same procedure to order more than two decimals from least to greatest or greatest to least.

EXAMPLE

▶ Order these decimals from greatest to least: 3.029, 3.061, 3.023, 3.148.

▶ Write the decimals in a column, lining up the decimal points.

3.029
3.061
3.023
3.148

▶ Order the numbers based on the first place where the digits differ.

     3.02**3**

     3.02**9**

     3.0**61**

     3.**148**

▶ In order from greatest to least the numbers are: 3.148 > 3.061 > 3.029 > 3.023.

Rounding Decimals

Rounding decimals involves the same principles as rounding whole numbers. The most important thing to keep in mind is the place value that we are rounding to. For example, what is 8.4739 rounded to the nearest tenth? To round the number, we must look at the next digit in the hundredths place. Since 7 is greater than 5, we round the tenths digit up. So, 8.4739 rounded to the nearest tenth is 8.5. Likewise, 8.4739 rounded to the nearest hundredth would be 8.47, since 3 is less than 5. A problem may ask us to round a sum, a difference, a product, or a quotient.

EXAMPLE

▶ What is 39.5748 rounded to the nearest thousandths?

▶ Identify the digit that represents the place value you are finding. Then, check the digit to right of this place. Round up the digit you are looking for if the digit to its right is 5 or greater. Keep the digit you are looking for the same if the digit to its right is less than 5.

▶ The thousandths place is three places to the right of the decimal point. Therefore, you must look to the ten thousandths place. Since the digit in the ten thousandths place is 7, the thousandths place would be rounded up from 4 to 5.

▶ 39.5748 rounded to the nearest thousandths place is 39.575.

Adding Decimals

The addition of decimals follows the same rules as the addition of whole numbers. The first thing to do when adding two decimals is to line up the numbers according to the placement of their decimal points. This is a very important step, since it helps assure that we are adding numbers with the same place value. Starting from the left and moving right, add columns, regrouping when necessary.

EXAMPLE

▶ Find the sum of 6.049 and 12.87.

▶ Align the numbers based on the place value of the digits. Represent 12.87 as 12.870 in order to keep the number of digits in both decimals the same.

$$
\begin{array}{r}
1 \\
6.049 \\
+\,12.870 \\
\hline
18.919
\end{array}
$$

▶ Notice that for any column which sums to 10 or greater, the tens digits is carried over to the place to the left of the column.

▶ Therefore, the sum of 6.049 and 12.87 is 18.919.

Here's another example.

EXAMPLE

▶ What is the sum of 5.36 and 2.456?

▶ Write the decimals so that the place values of the digits line up based on the decimal points and add a 0 where necessary.

$$
\begin{array}{r}
1 \\
5.360 \\
+\,2.456 \\
\hline
7.816
\end{array}
$$

The same vertical format can be used to add three or more decimals. Just remember to keep the number of places in each decimal the same.

EXAMPLE

▶ What is the sum of 4.37, 2.955, and 3.12?

▶ Write the decimals so that the place values of the digits line up based on the decimal points. Add 0s where they are needed.

$$
\begin{array}{r}
1\,1 \\
4.370 \\
2.955 \\
+\,3.120 \\
\hline
10.445
\end{array}
$$

Subtracting Decimals

Like the addition of decimals, the same rules apply to the subtraction of decimals.

▶ What is the difference between 9.465 and 5.28?

▶ Write the decimals so that the place values of the digits line up based on the decimal points. Add 0s where they are needed.

$$\begin{array}{r} {\scriptstyle 3\,1} \\ 9.\cancel{4}65 \\ -\,5.280 \\ \hline 4.185 \end{array}$$

Multiplying Decimals

We multiply decimals just as we do whole numbers. When we have found the product of the numbers, we must also correctly place the decimal point:

$$\begin{array}{r} 6.13 \\ \times\ 0.32 \\ \hline 1226 \\ 1839 \\ \hline 19616 \end{array}$$ → Where do we place the decimal point?

To find out where to place the decimal point, look at the factors in the problem. Why? The number of decimal places in the product is equal to the sum of the number of decimal places in both factors. The two factors in the problem are 6.13 and 0.32. Each factor has two places after the decimal point. So, the product of the problem will have four decimal places:

$$\begin{array}{r} 6.13 \\ \times\ 0.32 \\ \hline 1226 \\ 1839 \\ \hline 1.9616 \end{array}$$ ← Count four places from the last digit moving right to left.

So, the product of 6.13 and 0.32 is 1.9616.

Sometimes, we will be asked to round the product of two decimals to a particular place value, as shown in the example below.

EXAMPLE

▶ What is the product of 11.35 and 2.125 rounded to the nearest tenths place?

▶ Find the product as if you are solving a multiplication problem involving whole numbers.

$$
\begin{array}{r}
11.35 \\
\times\ 2.125 \\
\hline
5675 \\
2270 \\
1135 \\
2270 \\
\hline
2411875
\end{array}
$$

▶ Find the sum of the number of decimal places in both factors. Count five places from the last digit moving right to left and place the decimal point: 24.11875.

▶ Round the product to the nearest tenths place; 24.11875 rounded to the nearest tenths place is 24.1.

▶ The product of 11.35 and 2.125, rounded to the nearest tenths place, is 24.1.

To solve some problems involving decimals, we may need to use more than one operation.

EXAMPLE

▶ At the local farmers market, Jasmine buys 3.5 pounds of red bell peppers at a price of $2.95 per pound. She also buys 4.25 pounds of fingerling potatoes at $2.40 per pound. How much did Jasmine spend to the nearest penny?

▶ Use multiplication to solve each problem without placing the decimal point.

Peppers	Potatoes
2.95	2.40
× 3.5	× 4.25
1475	1200
885	480
10325	960
	102000

▶ Add the number of decimal places in each number to determine where to place the decimal point. Then, mark where the decimal point goes. Round each product to the nearest cent.

Red Peppers: $10325 \rightarrow 10.325 \rightarrow \$10.33$

▶ There are two decimal places in 2.95 and one decimal place in 3.5. So, there are three decimal places in all.

Potatoes: $102000 \rightarrow 10.2000 \rightarrow \$10.20$

▶ There are two decimal places in 2.40 and two decimal places in 4.25. So, there are four decimal places in all.

▶ Add the costs of both purchases.

$\$10.33 + \$10.20 = \$20.53$

▶ Jasmine spent a total of $20.53 for the red peppers and potatoes.

Dividing Decimals

Dividing decimals by a whole number follows the same steps as long division of whole numbers, except for the placement of the decimal point in the quotient. So, where does the decimal point go? Where we place the decimal point in a division problem depends on where the decimal point appears in the dividend. Consider this problem.

EXAMPLE

▶ A group of friends bought 8 tickets for a Saturday concert of their favorite group. The tickets cost a total of $319.60. How much did each ticket cost?

▶ Write the problem in long division format.

$$8\overline{)319.60}$$

▶ Place the decimal point in the quotient in the same position it appears in the dividend then solve.

$$
\begin{array}{r}
39.95 \\
8\overline{)319.60} \\
-24 \\
\hline
79 \\
-72 \\
\hline
76 \\
-72 \\
\hline
40 \\
-40 \\
\hline
0
\end{array}
$$

▶ Each concert ticket costs $39.95.

In the previous example, a decimal was divided by a whole number. In some problems, both the divisor and the dividend are decimals. In problems like these, we may need to add 0s and be especially careful about the placement of the decimal point in the quotient.

▶ Jenny pays $19.04 for 3.2 pounds of beads at a local craft store. What is the cost per pound of beads?

▶ Write the problem in long division format.

$$3.2\overline{)19.04}$$

▶ Multiply the divisor by a power of ten, so it is a whole number. Then, multiply the dividend by the same power of ten. In this problem, multiply both the divisor and dividend by 10.

▶ Think: How many places to the right must the decimal point be moved to make the divisor a whole number? Then, multiply the dividend by the same power.

$$3.2\overline{)19.04}$$

▶ Find the quotient by using long division. Add a 0 at the end of the dividend.

$$
\begin{array}{r}
5.95 \\
32\overline{)190.40} \\
-160 \\
\hline
304 \\
-288 \\
\hline
160 \\
-160 \\
\hline
0
\end{array}
$$

▶ The beads Jenny bought cost $5.95 per pound.

EXERCISES

EXERCISE 2–1

Answer each question using the figure provided.

1. What fraction does the shaded part of the image represent?

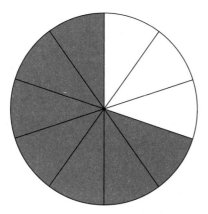

2. What fraction does the shaded part of the image represent?

3. What fraction does the shaded part of the image represent?

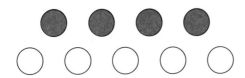

4. What fraction does the shaded part of the image represent?

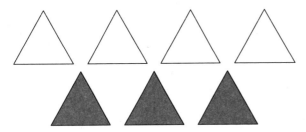

EXERCISE 2–2

Find two equivalent fractions for each fraction below.

1. $\dfrac{3}{7}$

2. $\dfrac{5}{12}$

3. $\dfrac{11}{25}$

4. $\dfrac{23}{40}$

EXERCISE 2-3

Write each fraction in simplest form.

1. $\dfrac{12}{18}$

2. $\dfrac{18}{30}$

3. $\dfrac{7}{28}$

4. $\dfrac{24}{72}$

EXERCISE 2-4

Write the answer to the following addition or subtraction problems in simplest form.

1. $\dfrac{12}{20} + \dfrac{6}{20}$

2. $\dfrac{13}{18} - \dfrac{7}{18}$

3. $\dfrac{3}{7} + \dfrac{5}{21}$

4. $\dfrac{9}{12} - \dfrac{1}{4}$

EXERCISE 2-5

Write the answer to the following addition or subtraction problems of mixed numbers in simplest form.

1. $3\dfrac{1}{3} + 2\dfrac{3}{5}$

2. $7\dfrac{5}{9} - 2\dfrac{2}{3}$

3. $5\dfrac{2}{3} + 3\dfrac{4}{9}$

4. $8\dfrac{7}{10} - 2\dfrac{13}{15}$

EXERCISE 2-6

Write the product of each multiplication problem in simplest form.

1. $\dfrac{3}{4} \times \dfrac{4}{5}$

2. $\dfrac{5}{8} \times \dfrac{3}{5}$

3. $3\dfrac{2}{3} \times 5\dfrac{3}{4}$

4. $1\dfrac{2}{5} \times 4\dfrac{3}{7}$

EXERCISE 2-7

Write the quotient of each of the following division problems in simplest form.

1. $\dfrac{4}{9} \div \dfrac{2}{3}$

2. $\dfrac{3}{4} \div \dfrac{1}{8}$

3. $2\dfrac{5}{6} \div \dfrac{1}{3}$

4. $3\dfrac{1}{2} \div 1\dfrac{2}{5}$

EXERCISE 2–8

Solve each of the following questions and write the answer as a fraction or mixed number in simplest form.

1. On Saturday, $\frac{2}{3}$ inch of rain fell. On Sunday, $\frac{3}{5}$ inch of rain fell. How much rain fell on Saturday and Sunday?

2. A stone weighs $6\frac{1}{5}$ ounces and another stone weighs $4\frac{2}{3}$ ounces. How much more does the heavier stone weigh than the lighter stone?

3. A recipe calls for $1\frac{1}{2}$ cup of milk. If you are making $3\frac{1}{2}$ times the recipe, how much milk will you need?

4. A gerbil needs $10\frac{1}{2}$ grams of food a day. If a bag of gerbil feed contains 80 grams, how many full days will it feed the gerbil?

EXERCISE 2–9

Write the decimal form of each number below.

1. fifty-two hundredths

2. thirty-nine and nine tenths

3. two hundred and forty-one and one hundred and twenty-five thousandths

4. fifteen ten thousandths

EXERCISE 2–10

Write the word name for each decimal.

1. 15.2

2. 21.45

3. 0.004

4. 5.0141

EXERCISE 2–11

Compare each pair of decimals using $>$, $<$, or $=$.

1. 1.249 _?_ 1.247

2. 5.170 _?_ 5.17

3. 33.75 _?_ 37.35

4. 121.532 _?_ 121.503

EXERCISE 2–12

Find the sum or difference for each addition and subtraction problem.

1. $3.15 + 2.44$

2. $6.52 + 3.185$

3. $7.635 - 4.18$

4. $32.47 - 26.63$

EXERCISE 2–13

Find the product or quotient for each of the following pairs of numbers.

1. 2.8×3.5

2. 5.75×6.32

3. $15.84 \div 0.4$

4. $26.68 \div 2.5$

EXERCISE 2–14

Solve the following problems and round each product or quotient to the nearest hundredth.

1. Sasha purchased 3.5 pounds of seedless grapes at $2.45 per pound. How much did the grapes cost?

2. If a box of a dozen popsicles cost $4.25, what is the average price for each popsicle?

3. A group of eight friends went to a baseball game and sat together. The on-line tickets they bought cost $156.40 including a service charge. What was the average price per ticket?

4. Jessica bought 2.5 pounds of cookies at a cost of $12.99 per pound. How much did Jessica pay for the cookies?

Flashcard App

Exponents, Roots, and Scientific Notation

e hear the word *power* used in everyday speech in sentences such as "The president has the power to veto the bill even though it was passed by Congress." In sports, the word is often used to refer to the force and speed displayed by an athlete. In mathematics, power also has a special meaning.

Powers and Exponents

A **power** is a number that can be written as the product of equal factors. In other words, a power is used to show repeated multiplication. Expressions such as 2^3 or 3^4 are the same as writing $2 \times 2 \times 2$ or $3 \times 3 \times 3 \times 3$. The number that is the factor in a power is called the **base**, and the number of times it is multiplied is the **exponent**. In 2^3, 2 is the base and 3 is the exponent. A number written as a base and exponent is said to be written in **exponential form**:

$$2^3 = 2 \times 2 \times 2 = 8$$
$$3^4 = 3 \times 3 \times 3 \times 3 = 81$$

A number multiplied by itself—that is, with the exponent 2—is **squared**, and a number multiplied by itself two times is **cubed**. The mathematical expression 2^3 can be read as "2 to the power of 3" or "2 to the third." Of course, there really is no limit as to how great an exponent can be.

▶ Write $5 \times 5 \times 5 \times 5$ in exponential form.

▶ Write the factor being multiplied as the base of the number: 5.

▶ Write the number of times the factor is being multiplied as the exponent: 4.

▶ $5 \times 5 \times 5 \times 5$ in exponential form is 5^4.

The **standard form** of a number written in exponential form is the value of the number. Using the previous example, 5^4 written in standard form is 625.

EXAMPLE

▶ Write 3^4 in standard form.

▶ Write the base as a factor in a multiplication problem as many times as indicated by the exponent.

$3 \times 3 \times 3 \times 3$

▶ Solve the multiplication problem.

$3 \times 3 \times 3 \times 3 = 81$

▶ 3^4 in standard form is 81.

Powers of 0 and 1

There are a few important details to keep in mind when working with powers. First, any number raised to the first power, such as 4^1 or 10^1 or 100^1, is equal to the number itself. Therefore, 4^1 equals 4, 10^1 equals 10, and 100^1 equals 100. The second important fact is that any *nonzero* number raised to the power of 0 equals 1. This may seem strange at first, but it's an important rule to keep in mind. Thus, 1^0, 10^0, 100^0, and $1,000,000^0$ all equal 1.

EXAMPLE

▶ Write 12^0 in standard form.

▶ Any number raised to the power of 0 equals 1.

▶ So, 12^0 equals 1.

Here's another problem that involves numbers raised to the power of 0.

▶ What is the product of 5^0 and 8^0?

▶ Find the value of each factor.

$$5^0 = 1$$
$$8^0 = 1$$

▶ Write the multiplication sentence and solve.

$$1 \times 1 = 1$$

▶ The product of 5^0 and 8^0 equals 1.

Negative Exponents

Not all exponents are 0 or positive. It's not uncommon to see expressions such as 2^{-3} or 3^{-2}, especially when we are simplifying more complex expressions. How do we go about evaluating them? Fortunately, the process is quite simple: Write the expression as a reciprocal with a positive power in the denominator and make the numerator 1. Therefore, $3^{-2} = \dfrac{1}{3^2}$. Then, simplify the expression. When we "evaluate" an expression such as 3^{-2}, we find its numerical value.

▶ Find the value of 5^{-3}.

▶ Write the reciprocal of the expression with a positive exponent in the denominator.

$$5^{-3} = \frac{1}{5^3}$$

▶ Find the value of the new expression:

$$5^{-3} = \frac{1}{5^3} = \frac{1}{5 \times 5 \times 5} = \frac{1}{125}$$

▶ The expression 5^{-3} equals $\dfrac{1}{125}$.

Raising Negative Numbers to a Positive Power

When a negative number is raised to a positive power that is even, then the resulting answer is positive or negative depending on whether the exponent is even or odd. If the exponent is even, then the product is positive.

EXAMPLE

▶ What is the value of the expression $(-5)^4$?

▶ Expand the expression.

$(-5) \times (-5) \times (-5) \times (-5)$

▶ Evaluate the new expression.

$(25) \times (25) = 625$

▶ The expression $(-5)^4$ equals 625.

Pay attention to the sign of the product in the example that follows.

▶ Evaluate $(-2)^5$.

▶ Write the terms of the factors of the problem and keep track of the signs.

$$(-2) \times (-2) \times (-2) \times (-2) \times (-2) = ?$$

▶ Find the value of the new expression by multiplying the terms in sequence.

$$(-2) \times (-2) \times (-2) \times (-2) \times (-2) = -32$$

▶ The expression $(-2)^5$ equals -32.

Notice that the solution to the preceding example is -32. Problems like this one ask us to multiply a negative number by itself. The sign of the product depends on how many times the factor is multiplied. It's easy to keep track of which sign the product will have by remembering the following rules: If a negative factor is multiplied an *even* number of times, the product is positive. However, if a negative factor is multiplied an *odd* number of times, the product is negative.

As the digit of the negative exponent increases, the value of the expression becomes smaller. Why does this happen? If we visualize a number line, we know that as we move from right to left the numbers become smaller. In the case of $(5)^{-3}$ and $(5)^{-4}$, $\dfrac{1}{625}$ is less than $\dfrac{1}{125}$.

On the following page is a convenient chart that shows some of the key exponent powers of ten.

$$
\begin{aligned}
10{,}000 &= 10^4 \\
1{,}000 &= 10^3 \\
100 &= 10^2 \\
10 &= 10^1 \\
1 &= 10^0 \\
0.1 &= 10^{-1} \\
0.01 &= 10^{-2} \\
0.001 &= 10^{-3} \\
0.0001 &= 10^{-4}
\end{aligned}
$$

 IRL Very large numbers sometimes have special names such as *googol*. The googol was first named by the American mathematician and author Edward Kasner. It represents 10^{100}, which is 1 followed by one hundred 0s. Kasner asked his 9-year-old nephew what he should call such a large number and the boy replied "a googol."

Properties of Exponents

As we've seen, a number often is much easier to work with when it is expressed in exponential form than when it is written in standard form. Writing and evaluating 10^8 is a lot easier than dealing with all the digits in 100,000,000. We can use exponential form along with several important rules, or **laws of exponents,** when we perform operations written in exponential form.

Let's consider how we would go about finding the product of 2^3 and 2^4. We could write out all the terms: $2 \times 2 \times 2 \times 2 \times 2 \times 2 \times 2 = 2^7$. Notice that the number of the terms equals the sum of the exponents of the original two expressions. The **product rule of exponents** states that, to multiply numbers with the same base, add the exponents.

▶ What is the product of 3^2 and 3^3, written in exponential form?

▶ Note the exponents in each expression.

$$3^2 \rightarrow \text{exponent } 2$$
$$3^3 \rightarrow \text{exponent } 3$$

▶ Add the exponents: $2 + 3 = 5$.

▶ Use the sum as the exponent of the product, which gives 3^5.

Another important rule tells us how to divide expressions with exponents. Recall that multiplication and division are inverse operations. The **quotient rule of exponents** states that to divide numbers with the same base, subtract the exponents.

▶ What is the quotient of 5^5 divided by 5^3, written in exponential form?

▶ Note the exponents in each expression.

$$5^5 \rightarrow \text{exponent } 5$$
$$5^3 \rightarrow \text{exponent } 3$$

▶ Subtract the exponents: $5 - 3 = 2$.

▶ Use the difference as the exponent of the quotient: 5^2.

▶ 5^5 divided by 5^3, or $\dfrac{5^5}{5^3}$, equals 5^2.

Some exponential expressions can themselves be raised to a power. For example, how would we go about evaluating the expression $(2^3)^2$? To solve the problem, we need to use the **power rule of exponents**, which states that to raise an exponential expression to a power, we must multiply the exponents.

EXAMPLE

▶ What is $(2^3)^2$?

▶ Note the exponents.

$$2^3 \rightarrow \text{exponent 3}$$
$$(2^3)^2 \rightarrow \text{exponent 2}$$

▶ Multiply the exponents: $3 \times 2 = 6$.

▶ Use the product of the exponents as the power of the base: 2^6.

$$(2^3)^2 = 2^3 \times 2^2 = 2^6 = 2 \times 2 \times 2 \times 2 \times 2 \times 2 = 64$$

▶ $(2^3)^2$ equals 2^6, or 64.

The rules of exponents are helpful when answering word problems.

EXAMPLE

▶ Suppose a certain bacterium doubles itself every hour. If a science experiment begins with 2 bacteria, how many will there be at the end of 10 hours? Write the answer in exponential form and then evaluate the expression to find the exact number of bacteria.

▶ Think about what the problem says.

The experiment begins with 2 bacteria.
The experiment goes on for 10 hours.

▶ Think about what the problem asks.

How many bacteria will there be after 10 hours?

▶ Write the number of bacteria at the start of the experiment using exponents: $2^1 = 2$.

▶ Write the expression that shows the number of bacteria after 10 hours as a power of a power: $(2^1)^{10} = 2^{10}$.

▶ Find the value of the expression 2^{10}.

$$2^{10} = 2 \times 2 \times 2 \times 2 \times 2 \times 2 \times 2 \times 2 \times 2 \times 2 = 1{,}024$$

▶ At the end of 10 hours, there will be 2^{10}, or 1,024, bacteria.

Roots

A **perfect square** is the product of an integer being multiplied times itself. In the 10-by-10 multiplication table below, all the highlighted numbers are examples of perfect squares:

×	1	2	3	4	5	6	7	8	9	10
1	1	2	3	4	5	6	7	8	9	10
2	2	4	6	8	10	12	14	16	18	20
3	3	6	9	12	15	18	21	24	27	30
4	4	8	12	16	20	24	28	32	36	40
5	5	10	15	20	25	30	35	40	45	50
6	6	12	18	24	30	36	42	48	54	60
7	7	14	21	28	35	42	49	56	63	70
8	8	16	24	32	40	48	56	64	72	80
9	9	18	27	36	45	54	63	72	81	90
10	10	20	30	40	50	60	70	80	90	100

Notice that all the perfect squares fall on the diagonal of this table, starting with the lowest number at the top left corner and the greatest number at the bottom right corner.

No matter how much we might extend the table—20 by 20, 50 by 50, 100 by 100, 1,000 by 1,000, and so on—the diagonal pattern continues: All the perfect squares fall on the diagonal.

Knowing how perfect squares are formed helps us to understand the concept of a square root. The **square root** of a number n is the number that produces n when multiplied by itself. In other words, perfect squares and square roots are inverse operations. Square roots are written under a **radical sign**:

$$\sqrt{100}$$

The number under the radical sign—the one you want to find the square root of—is called the **radicand**.

Every rational number has two square roots—one positive and one negative. Thus, the square roots of $\sqrt{9}$ are 3 and −3, since 3 times 3 equals 9 and −3 times −3 equals 9. The square roots of perfect squares are often written with a ± sign: $\sqrt{9} = \pm 3$. The positive square root of a number is called the **principal square root**. In this book, when we use the term *square root*, it refers only to the principal (positive) square root.

What if the number under the radical sign is *not* a perfect square? A convenient method for dealing with the square roots of nonperfect squares is to factor the number under the square sign so that we can pull out the root of any factor that is a perfect square. Let's look at an example.

EXAMPLE

▶ What is $\sqrt{75}$ in its simplest radical form?

▶ Identify the radicand, or number under the radical sign. Find the factors of the radicand so that one is a perfect square.

$$\sqrt{75} = \sqrt{25 \times 3}$$

▶ Find the principal square root of the perfect square.

$$\sqrt{25} = 5$$

▶ Write the square root outside the radical sign and leave the other factor under the radical sign.

$$\sqrt{75} = 5\sqrt{3}$$

▶ $\sqrt{75}$ in its simplest radical form is $5\sqrt{3}$.

The square root of any rational number that's not a perfect square, like $\sqrt{3}$ from our example, is an **irrational number**. Without a calculator, it would be very difficult to find the value of $\sqrt{3}$, or any other irrational number, so that's why we leave these numbers under the radical sign. With a calculator, though, it's easy to get an approximate value for $\sqrt{3}$, which is 1.7.

Estimating Roots

To find the value of an irrational root without a calculator, we can estimate based on what we know about perfect squares. Here's an example.

EXAMPLE

▶ What is an estimate of $\sqrt{43}$?

▶ Find the square root of perfect squares close to the number that is the radicand.

$$\sqrt{36} = 6 \text{ and } \sqrt{49} = 7$$

▶ Therefore, $\sqrt{43}$ will be somewhere in between 6 and 7.

▶ Estimate the location of the radicand between the perfect square numbers. 43 is about halfway between 36 and 49, so $\sqrt{43}$ should be about halfway between 6 and 7.

$$\sqrt{43} \approx 6.5$$

▶ Check your estimate with a calculator. If you input $\sqrt{43}$ into a calculator, you get 6.5574..., so the estimate of 6.5 is very close.

Adding and Subtracting Roots

Just as we can add expressions with the same base and exponent, we can also directly add or subtract radicals that have the same root under the radical sign: $2\sqrt{5} + 7\sqrt{5} = 9\sqrt{5}$ and $8\sqrt{3} - 6\sqrt{3} = 2\sqrt{3}$. With unlike radical expressions—that is, those with different radicands—we may be able to simplify one or more of the radical expressions, which may make addition or subtraction possible.

EXAMPLE

▶ What is the sum of $3\sqrt{12}$ and $2\sqrt{27}$?

▶ Simplify one or both of the radical expressions so they express a perfect square and the same radicand.

$$3\sqrt{12} = 3\sqrt{4 \times 3}$$

▶ Pull out the square root of 4, which is 2. Multiply the coefficient 3 that is already outside the radicand by 2.

$$3\sqrt{4 \times 3} = 2 \times 3\sqrt{3} = 6\sqrt{3}$$

▶ Simplify $2\sqrt{27}$ so it has a like radicand: $\sqrt{3}$.

$$2\sqrt{27} = 2\sqrt{3 \times 9} = 3 \times 2\sqrt{3} = 6\sqrt{3}$$

▶ Finally, add the two like radical expressions.

$$6\sqrt{3} + 6\sqrt{3} = 12\sqrt{3}$$

▶ The sum of $3\sqrt{12}$ and $2\sqrt{27}$ is $12\sqrt{3}$.

Let's look at another example involving unlike radical expressions. Although this example involves subtraction, it uses the same basic skills.

▶ What is the difference between $5\sqrt{54}$ and $3\sqrt{24}$?

▶ Simplify one or both of the radical expressions so they express a perfect square and the same radicand.

$$5\sqrt{54} = 5\sqrt{9 \times 6} = 3 \times 5\sqrt{6} = 15\sqrt{6}$$
$$5\sqrt{24} = 3\sqrt{4 \times 6} = 2 \times 3\sqrt{6} = 6\sqrt{6}$$

▶ Subtract the two radical expressions.

$$15\sqrt{6} - 6\sqrt{6} = 9\sqrt{6}$$

▶ The difference between $5\sqrt{54}$ and $3\sqrt{24}$ is $9\sqrt{6}$.

Multiplying Roots

Just as with exponents, we can multiply radical expressions easily. If the radicals have no coefficients, we multiply the radicands and, if we can, simplify the product.

EXAMPLE

▶ What is the product of $\sqrt{12}$ and $\sqrt{6}$?

▶ Write the radicals as a multiplication sentence.

$$\sqrt{12} \times \sqrt{6} = ?$$

▶ Multiply the radicands.

$$\sqrt{12} \times \sqrt{6} = \sqrt{72}$$

▶ Factor the new radicand looking for a perfect square as a factor.

$$\sqrt{72} = \sqrt{36 \times 2}$$

▶ Simplify the expression. Since $\sqrt{36}$ equals 6, we can pull that out and place it before the radical sign.

$$\sqrt{36 \times 2} = 6\sqrt{2}$$

▶ The product of $\sqrt{12}$ and $\sqrt{6}$ is $6\sqrt{2}$.

If the radical expressions have coefficients, we must multiply the coefficients first, then multiply the radicands, and (if possible) simplify the expression.

EXAMPLE

▶ What is the product of $4\sqrt{8}$ and $3\sqrt{5}$?

▶ Multiply the coefficients, and then multiply the radicands.

$$4 \times 3 = 12$$
$$\sqrt{8} \times \sqrt{5} = \sqrt{40}$$

▶ Write the product as a radical expression, and simplify.

$$12\sqrt{40} = 12\sqrt{4 \times 10} = 24\sqrt{10}$$

▶ The product of $4\sqrt{8}$ and $3\sqrt{5}$ is $24\sqrt{10}$.

The product of two radical expressions is not always another radical expression. Sometimes, the answer is a rational number.

▶ What is the product of $3\sqrt{8}$ and $3\sqrt{2}$?

▶ Multiply the coefficients, and then multiply the radicands.

$$3 \times 3 = 9$$
$$\sqrt{8} \times \sqrt{2} = \sqrt{16}$$

▶ Write product as a radical expression, and simplify.

$$9\sqrt{16} = 9 \times 4 = 36$$

▶ The product of $3\sqrt{8}$ and $3\sqrt{2}$ is 36.

Dividing Roots

To divide radical expressions, we must think of the two roots as a fraction that we will simplify.

▶ What is the quotient of $10\sqrt{12}$ divided by $2\sqrt{3}$?

▶ First, we must set up the problem properly by factoring each expression according to its coefficients and its radicals.

$$10\sqrt{12} \div 2\sqrt{3} = \frac{10}{2} \times \sqrt{\frac{12}{3}}$$

▶ Next, we simplify each factor and the resulting expression.

$$\frac{10}{2} = 5 \text{ and } \sqrt{\frac{12}{3}} = \sqrt{4}$$

$$\rightarrow \frac{10}{2} \times \sqrt{\frac{12}{3}} = 5\sqrt{4} = 5 \times 2 = 10$$

▶ The quotient of $10\sqrt{12}$ divided by $2\sqrt{3}$ is 10.

What happens when one of the parts of the radical expressions does not divide evenly? In the case of coefficients, we just leave the quotient as a fraction:

$$5\sqrt{2} \div 3\sqrt{2} = \frac{5}{3} \times \frac{\sqrt{2}}{\sqrt{2}} = \frac{5}{3} \times 1 = \frac{5}{3}$$

When roots do not divide evenly, we are required to take an extra step since the rules of mathematics do not allow a radical to appear in the denominator. (This is a hard and fast rule, much like the rule that you cannot have a 0 in the denominator of a fraction.) Fortunately, we can always eliminate these radicals by rationalizing the denominator, that is, by converting a denominator with a root into an integer.

EXAMPLE

▶ What is the quotient of $9\sqrt{5}$ divided by $3\sqrt{2}$?

▶ Set up the problem so the coefficients and roots can be divided separately.

$$9\sqrt{5} \div 3\sqrt{2} = \frac{9}{3} \times \frac{\sqrt{5}}{\sqrt{2}} = 3\frac{\sqrt{5}}{\sqrt{2}}$$

▶ To clear $\sqrt{2}$ from the denominator, multiply the expression by $\dfrac{\sqrt{2}}{\sqrt{2}}$.

Remember that any number with the same numerator and denominator equals 1.

$$3\frac{\sqrt{5}}{\sqrt{2}} \times \frac{\sqrt{2}}{\sqrt{2}} = 3\frac{\sqrt{10}}{\sqrt{4}} = 3\frac{\sqrt{10}}{2}$$

▶ The quotient of $9\sqrt{5}$ divided by $3\sqrt{2}$ is $3\dfrac{\sqrt{10}}{2}$.

Scientific Notation

Scientists deal with a wide range of numbers, from the tiniest subatomic particle to the vast stretches of space. Scientific notation helps scientists to perform calculations with such extreme measures.

A number written in **scientific notation** has the form of $c \times 10^n$. The c stands for a constant that is a decimal number greater than 1 and less than 10. The other factor, 10^n, represents a power of ten. For example, by some estimates the Milky Way galaxy contains about 400 billion stars. Written in standard notation, this is 400,000,000,000. To write this number in scientific notation, we first find the beginning number that is between 1 and 10 and then we multiply it by 10^n. In this example, the number of decimal places is 11. We can calculate this by moving from right to left 11 places:

400,000,000,000

In scientific notation, the 400,000,000,000 stars in the Milky Way $=$ 4.0×10^{11}.

Let's look at an example together.

EXAMPLE

▶ The distance from Earth to the moon is approximately 239,000 miles. What is this distance written in scientific notation?

▶ Examine the beginning digits to find a decimal greater than 1 and less than 10. Form the decimal by moving five places from right to left.

$$239,000 \rightarrow 2.39$$

▶ 2.39 is a decimal greater than 1 and less than 10. Write the number as the product of the decimal and the power of ten: 2.39×10^5.

▶ The distance of 239,000 miles from Earth to the moon is 2.39×10^5 miles.

To write a number expressed in scientific notation in standard form, we reverse the steps.

EXAMPLE

▶ The distance from Earth to Mars is approximately 3.48×10^7 miles. What is this distance written in standard form?

▶ Write the value of the power of ten in standard form. To do this, write 1 and then add the number of 0s based on the exponent: $10^7 = 10,000,000$.

▶ Find the product of the decimal and the power of ten written in standard form.

$$3.48 \times 10,000,000 = 34,800,000$$

▶ The distance from Earth to Mars is approximately 34,800,000 miles.

BTW

When you write a power of ten in standard form, remember to begin with 1—not 10—and follow it with the number of 0s named by the exponent.

Earlier in this chapter, we saw that numbers written with negative exponents are small and they get smaller as the negative exponent gets larger. These facts help us to write very small numbers. Let's consider an example.

EXAMPLE

▶ How do we write 0.0045 in scientific notation?

▶ Move right from the decimal point until we form a number greater than 1 and less than 10.

> 0.0045 → 4.5

▶ Write the number of places the decimal point was moved as a power of ten using a negative exponent.

> $10^{-3} = 0.001$

▶ Write the product of the decimal times the power of ten.

> $0.0045 = 4.5 \times 0.001 = 4.5 \times 10^{-3}$

▶ 0.0045 written in scientific notation is 4.5×10^{-3}.

Now let's take a look at how two numbers with different negative exponents compare.

EXAMPLE

▶ Which is larger, the width of an oxygen molecule at 6.0×10^{-11} or the width of a carbon dioxide molecule at 5.0×10^{-10}?

▶ We can solve this problem without calculations by examining the exponents. 10^{-10} is ten times larger than 10^{-11}. Alternately, 10^{-11} is ten times smaller than 10^{-10}.

▶ Therefore, the width of a molecule of carbon dioxide is greater than the width of a molecule of oxygen.

▶ We can check this answer by writing each number in standard form and then comparing them.

$$6 \times 10^{-11} = 6 \times 0.00000000001 = 0.00000000006$$
(oxygen molecule)
$$5 \times 10^{-10} = 5 \times 0.0000000001 = 0.0000000005$$
(carbon dioxide molecule)

▶ Since 0.0000000005 is greater than 0.00000000006, the answer is correct. The width of a molecule of carbon dioxide is, in fact, larger than the width of a molecule of oxygen.

Here's another example that involves making a comparison using scientific notation, but in this example we must also use some basic operations.

EXAMPLE

▶ The mass of Saturn is estimated to be 5.68×10^{26} kilograms, and the mass of Pluto is estimated at 1.31×10^{22} kilograms. In scientific notation, about how many times Pluto's mass is Saturn's mass?

▶ Find the quotient of the decimals of the two planets.

$$\frac{\text{Saturn's mass}}{\text{Pluto's mass}} = \frac{5.68}{1.31} = 4.34$$

▶ Find the quotient of the powers of ten.

$$\frac{\text{Saturn's mass}}{\text{Pluto's mass}} = \frac{10^{26}}{10^{22}} = 10^4$$

▶ Write the product of the two quotients in scientific notation.

$$4.34 \times 10^4$$

▶ The mass of Saturn is about 4.34×10^4 times greater than the mass of Pluto.

EXERCISES

EXERCISE 3–1

Write each number in exponential form.

1. $3 \times 3 \times 3$

2. $2 \times 2 \times 2 \times 2$

3. $4 \times 4 \times 4 \times 4 \times 4$

4. $10 \times 10 \times 10 \times 10 \times 10 \times 10$

EXERCISE 3–2

Write each number in standard form.

1. 2^4

2. 5^3

3. 8^1

4. 10^6

5. $3^3 \times 100^0$

EXERCISE 3–3

Evaluate each expression with a negative exponent or negative number and write it in standard form.

1. 2^{-3}

2. 6^{-2}

3. $(-3)^5$

4. $(-4)^4$

EXERCISE 3–4

Find the product of each multiplication problem and write the product in standard form.

1. $5^2 \times 5^4$

2. $2^5 \times 2^3$

3. $3^4 \times 3^3$

4. $7^2 \times 7^0$

EXERCISE 3–5

Solve each division problem and write the quotient in standard form.

1. $3^6 \div 3^2$

2. $2^7 \div 2^3$

3. $5^7 \div 5^2$

4. $3^1 \div 10^0$

EXERCISE 3–6

Find the value of the power of a power for each expression that follows. Write the final value in standard form.

1. $(3^2)^3$

2. $(2^4)^2$

3. $(7^2)^2$

4. $(10^2)^4$

EXERCISE 3-7

Answer the following questions involving radical expressions.

1. Reduce $\sqrt{48}$ as much as possible.

2. Approximate the value of $\sqrt{51}$.

3. What is $3.5\sqrt{6}$ plus $7.8\sqrt{6}$?

4. What is $5.78\sqrt{3}$ minus $6.26\sqrt{3}$?

5. What is $\sqrt{32}$ times $2.9\sqrt{2}$?

6. What is $12\sqrt{7}$ divided by $3\sqrt{5}$?

EXERCISE 3-8

Write the numbers in 1 and 2 in standard form.

1. 2.4×10^5

2. 4.1×10^8

Write the numbers in 3 and 4 in scientific notation.

3. 0.00073

4. 0.0000025

EXERCISE 3-9

Use what we've learned about scientific location to answer each question. Write the answer in scientific notation or standard form, depending on what the question indicates.

1. The distance between our Sun and Pluto is 5.91×10^{11} m. What is this distance written in standard form?

2. The distance between Earth and Venus is approximately 261,000,000 kilometers. What is this distance written in scientific notation?

3. Katy examines a human hair under an electron microscope. She finds that the hair is 0.0085 centimeters wide. What is the width of the hair written in scientific notation?

4. The width of an average red blood cell is approximately 0.000007 centimeters. What is the width of a red blood cell written in scientific notation?

Flashcard App

Equations and Inequalities

MUST KNOW

⚡ An equation is a mathematical statement that tells us that two expressions are equal by placing them on either side of an equal sign.

⚡ A function is a relationship in which each input number in one set is paired with exactly one output number in another set.

⚡ An inequality is a statement that tells us that two expressions are not necessarily equal by placing them on either side of an inequality symbol.

⚡ Equations and inequalities can be solved using basic operations or by graphing them on the coordinate plane.

Many students and adults think of algebra as an abstract subject that has little practical value to what happens in their daily lives. Nothing, however, could be further from the truth. In fact, we tend to use algebraic principles and operations many times a day. Consumers use algebra when they compare prices, evaluate discounts, and determine the value of interest rates. Engineers, chemists, and computer scientists use algebraic thinking and operations throughout the workday. Accountants, business managers, and banks use algebra to calculate profit and loss, exchange rates, depreciation, and returns on investments. Algebra even plays a role in sports as players use their understanding and skills to calculate what it will take to score points.

The Language of Algebra

When people think about algebra, the word that most frequently comes to mind is *equation*. An **equation** is a mathematical sentence formed by placing an equal sign between two expressions. Algebraic expressions involve variables, numbers, and operation symbols. In algebraic expressions, **terms** are separated by **operation symbols**. For example, the expression $2x^2$ has one term, whereas the expression $3x + 7$ has two terms, with the plus sign as the operation symbol. The **solution** of an equation is a number that we substitute for a variable, such as x, that makes the equation true.

For example, suppose you make $45 mowing and caring for lawns on a weekend. If your supplies, such as gasoline and lawn fertilizer, cost $13, you can find your profit with a simple equation: Lawn fees (F) − cost of supplies (S) = profit (P). In this case, the equation is: $F - S = P$, or $45 − $13 = $32.

An equation has specific values that provide a solution to the terms provided by the equation. A function is different. A **function** pairs each "input" number in one set with exactly one "output" number in another set. An example of a function is "each output number is $2.50 times the input number (the number of brownies)," which can be written as $y = 2.5x$. Suppose x equals the number of brownies and $2.50 equals the cost for each brownie. Using the function, we can easily calculate the cost of any number (y) of brownies purchased—4 brownies cost $10, 10 brownies cost $25, and so on.

Writing Expressions and Equations

In algebra, it's common to use a **variable** in the form of a letter such as $a, n, x,$ or y to represent an unknown number. Expressions that contain at least one variable are called **variable** expressions, for example, $a + 5, n - 7,$ $4 \times x,$ and $y \div 2.$ The meaning of each of these expressions is pretty obvious: the sum of a number plus 5, the difference between a number and 7, the product of a number times 4, and the quotient of a number divided by 2.

Certain words and phrases provide clues that indicate which operation is represented by a variable expression:

Addition	plus, added to, sum of, more than
Subtraction	minus, subtracted from, difference between, less than
Multiplication	times, multiplied by, product of
Division	divided by, quotient of, equal parts

EXAMPLE

▸ Write a variable expression that represents 8 more than a number, n.

▸ The phrase *more than* indicates that this expression involves addition.

▸ The variable expression $n + 8$ represents "8 more than a number, n."

Here's another example.

EXAMPLE

▶ Provide a variable expression that represents a number, n, multipled by 5.

▶ The phrase *multiplied by* indicates that this expression involves multiplication.

▶ The variable expression $n \times 5$, or $5n$, represents "a number, n, multiplied by 5."

Let's look at an example of how to translate an expression into an equation.

EXAMPLE

▶ How would you write "720 is the same as the product of 15 and x"?

▶ The word *product* indicates that this equation involves multiplication.

▶ The equation $15 \times x = 720$ or $15x = 720$ represents "720 is the same as the product of 15 and x."

Here's an example involving a different operation.

EXAMPLE

▶ How would you write "a is the quotient of t divided by 12"?

▶ *Quotient* and *divided by* indicate that this equation involves division.

▶ The equation $a = t \div 12$ or $a = \dfrac{t}{12}$ represents "a is the quotient of t divided by 12."

The following example shows how to translate an expression involving subtraction into an equation.

EXAMPLE

▶ How would you write "90 less than g equals x"?

▶ *Less than* indicates that this equation involves subtraction.

▶ The equation $x = g - 90$ represents "90 less than g equals x."

This final example shows how to translate an expression involving addition into an equation.

EXAMPLE

▶ What equation would you write to express "z is the same as the sum of of 22 and k"?

▶ The word *sum* indicates that this equation involves addition.

▶ The equation $z = 22 + k$ represents "z is the same as the sum of of 22 and k."

Solving Equations Using Addition or Subtraction

Many simple equations such as $x + 10 = 12$ and $x - 4 = 6$ can be solved using mental math. When solving equations, it's important to recall that addition and subtraction are **inverse operations**—opposite operations that undo each other. Our goal in solving equations is to isolate the unknown variable (often labeled x) on one side of the equation. For example, we know that 2 solves the equation $x + 10 = 12$, since 2 plus 10 equals 12. To find the solution to this equation, we must subtract 10 from both sides. Subtracting 10 "undoes" the addition in $x + 10$. An essential point to remember is that whatever we do to one side of an equation, we must do to the other side of the equation. The two equations $x + 10 = 12$ and $x = 2$ are called **equivalent equations**.

EXAMPLE

▶ What is the value of x in the equation $x + 6 = 9$?

▶ To solve for x, subtract 6 from both sides of the equation.

$$x + 6 = 9$$
$$x + 6 + -6 = 9 + -6$$
$$x + 0 = 3$$
$$x = 3$$

▶ Check your solution by substituting the value you found for x back into the original equation and then evaluating the result to determine if it makes a true mathematical statement.

$$x + 6 = 9$$
$$3 + 6 = 9$$
$$9 = 9 ✓$$

▶ Since both sides of the equation equal 9, the solution is correct. The checkmark (✓) after $9 = 9$ indicates that the statement is mathematically true and, therefore, "checks out." We will use this symbol in all relevant examples that follow.

Let's try an example that involves using addition to solve an equation.

EXAMPLE

▶ What is the value of y in the equation $y - 8.25 = 11.75$?

▶ To solve for y, add 8.25 to both sides of the equation.

$$y - 8.25 = 11.75$$
$$y - 8.25 + 8.25 = 11.75 + 8.25$$
$$y + 0 = 20$$
$$y = 20$$

▶ As we did in the preceding example, we check our solution by substituting the value we found for x back into the original equation and then evaluating the result to determine if it makes a true mathematical statement.

$$y - 8.25 = 11.75$$
$$20 - 8.25 = 11.75$$
$$11.75 = 11.75 \checkmark$$

▶ Since both sides of the equation equal 11.75, the solution is correct.

Addition, subtraction, or both operations may be necessary to solve equations based on word problems.

EXAMPLE

▶ Janice, Zaide, and Amy spent $87 for dinner. If Janice's dinner cost $27 and Zaide's dinner cost $31, how much did Amy's dinner cost?

▶ The cost of the three meals is the sum of three addends, so the problem should be expressed by an equation that involves addition.

$$\$27 + \$31 + x = \$87$$

▶ To solve for x, first simplify both sides of the equation and then solve.

$$\$27 + \$31 + x = \$87$$
$$\$58 + x = \$87$$
$$\$58 + (-\$58) + x = \$87 - \$58$$
$$x = \$29$$

▶ Once again, check the solution by substituting the value found for x back into the original equation. Then evaluate the equation to determine if it results in a true mathematical statement.

$$\$27 + \$31 + x = \$87$$
$$\$27 + \$31 + \$29 = \$87$$
$$\$87 = \$87 \checkmark$$

Solving Equations Using Multiplication or Division

Like addition and subtraction, multiplication and division are inverse operations. We can see that these operations "undo" each other by considering pairs of related equations, such as $12 \times x = 48$ and $48 \div 12 = x$.

EXAMPLE

▶ What is the value of x in the equation $\dfrac{x}{3} = 12$?

▶ Since the equation involves a fraction (that is, division), we must use the inverse operation of multiplication to solve for x. Therefore, multiply both sides of the equation by 3.

$$\frac{3}{1} \times \frac{x}{3} = \frac{3}{1} \times \frac{12}{1}$$

$$\frac{3}{3}x = 36$$

$$x = 36$$

▶ Check the solution by substituting the value for x into the original equation and then evaluating the equation.

$$\frac{x}{3} = 12$$

$$\frac{36}{3} = 12$$

$$12 = 12 \checkmark$$

As with multiplication, dividing both sides of an equation by the same (nonzero) number results in an equivalent equation.

EXAMPLE

▶ What is the value of x in the equation $1.5x = 21$?

▶ Since $1.5x$ is a multiplication expression, to solve for x, we must use the inverse operation and divide both sides of the equation by 1.5.

$$1.5x = 21$$

$$\frac{1.5x}{1.5} = \frac{21}{1.5}$$

$$x = 14$$

▶ Check the solution by substituting the value for x into the original equation and then evaluating the equation.

$$1.5x = 21$$

$$1.5(14) = 21$$

$$21 = 21 \checkmark$$

A combination of multiplication and division is often used to solve equations based on word problems.

▶ A store sells 15 bottles of hand sanitizer for a total of $82.50. Write an equation that represents the cost per bottle and then solve it.

▶ Use x to represent the cost of one bottle of hand sanitizer. Since we know the cost of 15 bottles, write the problem as an equation that uses multiplication.

$$15x = \$82.50$$

▶ To solve for x, use the inverse of multiplication. Divide both sides of the equation by 15 and then simplify.

$$15x = \$82.50$$

$$\frac{15x}{15} = \frac{\$82.50}{15}$$

$$x = \$5.50$$

▶ Check the solution by substituting the value for x into the original equation and evaluating.

$$15x = \$82.50$$

$$15 \times \$5.50 = \$82.50$$

$$\$82.50 = \$82.50 \checkmark$$

Equations with Fractions and Decimals

Since the product of reciprocals always equals 1, using them is helpful when clearing fractions from an equation and simplifying its solution.

▶ What is the value of x in the equation $\dfrac{2x}{5} = -20$?

▶ To solve for x, multiply both sides of the equation by $\dfrac{5}{2}$, which is the reciprocal of $\dfrac{2}{5}$.

$$\frac{2x}{5} = -20$$

$$\frac{5}{2} \times \frac{2x}{5} = \frac{5}{2} \times -20$$

$$x = \frac{-100}{2}$$

$$x = -50$$

▶ Check the solution by substituting the value for x into the original equation and then evaluating the equation.

$$\frac{2x}{5} = -20$$

$$\frac{2}{5} \times -50 = -20$$

$$\frac{-100}{5} = -20$$

$$-20 = -20 \checkmark$$

To clear an equation with decimals, we first need to multiply both sides by a power of ten, such as 10 or 100, depending on the place value of the decimal.

► What is the value of x in the equation $0.35x = 14$?

► To solve for x, first multiply both sides of the equation by 100 since the decimal is expressed as hundredths. Then divide both sides of the equation by 35 to solve for the value of x.

$$0.35x = 14$$
$$100 \times 0.35x = 100 \times 14$$
$$35x = 1400$$
$$\frac{35x}{35} = \frac{1400}{35}$$
$$x = 40$$

► Check the solution by substituting the value for x into the original equation and evaluating.

$$0.35x = 14$$
$$0.35 \times 40 = 14$$
$$14 = 14 \checkmark$$

Multi-Step Equations

Notice that in the previous example, both multiplication and division were used. It is very common to use more than one operation to solve an equation for x. Here's another example of a multi-step equation.

► What is the value of x in the equation $0.25x - 15 = -35$?

► To solve for x, first isolate the expression with x on one side of the equation.

$$0.25x - 15 = -35$$
$$0.25x - 15 + 15 = -35 + 15$$
$$0.25x = -20$$

▶ Next, divide each side by the decimal 0.25 to solve for the value of x.

$$\frac{0.25x}{0.25} = \frac{-20}{0.25}$$
$$x = -80$$

▶ Check the solution by substituting the value for x into the original equation and evaluating the equation.

$$0.25x - 15 = -35$$
$$(0.25 \times -80) - 15 = -35$$
$$-20 + (-15) = -35$$
$$-35 = -35 \checkmark$$

Inequalities

While an equation shows that two expressions have the same value, an **inequality** shows that two expressions are not equal. Instead of an equal sign, an inequality uses one of the following signs:

> $>$ greater than
> $<$ less than
> $\geq$ greater than or equal to
> $\leq$ less than or equal to

The same operations that are used to solve equations are used to solve inequalities.

▶ What is the solution of the inequality $x + 5 \leq 9$?

▶ Write the inequality.

$$x + 5 \leq 9$$

▶ Subtract 5 from both sides of the inequality.

$$x + 5 - 5 \leq 9 - 5$$

▶ Solve the inequality.

$$x \leq 4$$

▶ The solution to the inequality $x + 5 \leq 9$ is $x \leq 4$. This means that any number equal to or less than 4 is a solution.

▶ Check the solution by substituting one or more of the values for x into the original inequality and evaluating.

$x + 5 \leq 9$	$x + 5 \leq 9$
$4 + 5 \leq 9$	$2 + 5 \leq 9$
$9 \leq 9$ ✓	$7 \leq 9$ ✓

Combinations of operations—for example, addition and division or subtraction and multiplication—may be used to solve an inequality.

▶ What is the solution of the inequality $3x - 7 \geq 29$?

▶ Write the original inequality.

$$3x - 7 \geq 29$$

▶ Add 7 to both sides of the inequality and simplify.

$$3x - 7 + 7 \geq 29 + 7$$
$$3x \geq 36$$

▶ Divide both sides of the inequality by 3.

$$\frac{3x}{3} \geq \frac{36}{3}$$
$$x \geq 12$$

▶ So, the solution of the inequality $3x - 7 \geq 29$ is $x \geq 12$. This means that any number 12 or greater is a solution.

▶ Check the solution by substituting one or more of your values for x into the original inequality and evaluating.

$$3x - 7 \geq 29 \qquad\qquad 3x - 7 \geq 29$$
$$(3 \times 12) - 7 \geq 29 \qquad (3 \times 15) - 7 \geq 29$$
$$36 - 7 \geq 29 \qquad\qquad 45 - 7 \geq 29$$
$$29 \geq 29 \checkmark \qquad\qquad 38 \geq 29 \checkmark$$

 IRL   Businesses use inequalities to plan production activities, organize warehouses, and set prices. You may have come across inequalities in real life if you have ever taken an amusement park ride that says you must be 4 feet or taller to ride.

Another way to represent the solution of an inequality is by graphing it on a number line.

EXAMPLE

▶ What is the solution of the inequality $-2z < 6$? Graph the solution on a number line.

▶ Write the original inequality.

$$-2z < 6$$

▶ Divide both sides of the inequality by -2.

$$\frac{-2}{-2}z < \frac{6}{-2}$$

▶ Simplify and solve. Remember that whenever you multiply or divide an inequality by a negative number, you must reverse the sign.

$$\frac{-2}{-2}z < \frac{6}{-2}$$
$$z > -3$$

BTW

To show that -3 is not part of the solution, we use an open circle. If -3 were part of the solution, then we'd use a closed circle.

▶ The solution to the inequality $-2z < 6$ is $z > -3$.

▶ Graph the solution on a number line.

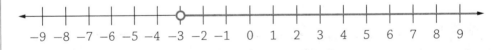

EXERCISES

EXERCISE 4–1

Write a variable expression or equation to represent each statement.

1. 4 less than a number w

2. a number n divided by 8

3. x is the same as the product of 4 and y

4. a is the sum of 24 and z

EXERCISE 4–2

Solve each equation for the unknown variable.

1. $w + 9 = 12$

2. $z - 4.75 = 10.5$

3. $\dfrac{x}{4} = -6$

4. $2.5y = 40$

EXERCISE 4–3

Solve each inequality for the unknown variable. Graph the solution on a number line.

1. $z + 9 \geq 12$

2. $5x - 14 < 26$

3. $y - 4 > 2$

4. $-\dfrac{1}{3}y + 5 < 7$

EXERCISE 4–4

Use what you've learned about algebraic equations to solve the following word problems.

1. The school dance committee spent $85 for paper streamers, balloons, and party mix. The paper streamers cost $15, and the balloons cost $12. How much did the party mix cost? Use c for total cost, s for the cost of streamers, b for the cost of balloons, and p for cost of party mix when writing an equation that represents the situation. Then, solve the equation.

2. At a gymnastics competition, Maya won points in five categories. The average of her five scores was 49. How many points did Maya earn at the competition? Write an equation to represent this situation and solve. Use x to represent Maya's total score.

3. In 2019, about 4.5 billion passengers boarded air flights across the globe. To the nearest thousand, what was the average number of flyers per day during this year? Write an equation to represent these facts. Use f to represent the average number of flyers per day.

4. It costs Ari $3 for a one-way bus trip to and from the local recreation center. A monthly bus pass costs $54. What is the fewest number of one-way bus trips Ari must take to make buying a monthly pass less costly than paying for each one-way trip? Write an inequality to represent this situation and solve.

5. Briana has a $50 gift card to a local plant nursery. She plans to buy impatiens plants for her garden that cost $3 per plant. She pays a $5 delivery charge for the nursery to bring the plants to her house. How many impatiens plants can Briana purchase without spending more than the amount of her gift card? Write an inequality to represent this situation and solve.

Flashcard App

Ratio, Proportion, and Percent

MUST ⚡ KNOW

⚡ A ratio compares two numbers using division.

⚡ Equivalent ratios are ratios that represent the same comparison using different numbers.

⚡ A rate is a special type of ratio that compares two related quantities measured in different units, such as miles per hour or dollars per pound.

⚡ A percent is a ratio that compares a number to 100.

atios, proportions, and percents are related concepts. Each makes a comparison between two numbers. In this chapter, we'll explore these concepts and make their connections clear.

Ratios

A **ratio** uses division to make a comparison between the sizes of two numbers. Ratios can be expressed in three different ways. For example, if the drama club has 5 boys and 7 girls, then the ratio of boys to girls in the drama club can be expressed as 5 to 7, 5:7, or $\frac{5}{7}$. In all three cases, the ratio is read aloud as "5 to 7." Sometimes, a ratio can be simplified. For example, if the number of boys to girls in the art club is 4 to 6, we can simplify $\frac{4}{6}$ as $\frac{2}{3}$, or "2 to 3."

EXAMPLE

▶ Olivia's vintage CD collection includes 6 hip hop CDs and 7 rock CDs. What is the ratio of hip hop CDs to rock CDs in Olivia's collection? Write the ratio in three different ways.

▶ Think of the ratio as 6 hip hop CDS to 7 rock CDs, or $\dfrac{6 \text{ hip hop CDs}}{7 \text{ rock CDs}}$.

▶ The ratio of hip hop to rock CDs is 6 to 7, 6:7, or $\frac{6}{7}$.

Here's a problem that involves finding a ratio in its simplest form.

EXAMPLE

The sophomore class in your high school has 56 girls and 40 boys. What is the ratio of girls to boys in the sophomore class? Write the ratio as a fraction in its simplest form.

▶ Think of the ratio as 56 girls to 40 boys, or $\dfrac{56 \text{ girls}}{40 \text{ boys}}$.

▶ Simplify the numerator and denominator.

$$\frac{56 \div 8}{40 \div 8} = \frac{7}{5}$$

▶ The ratio of girls to boys is $\dfrac{7}{5}$.

Equivalent Ratios

Two ratios that have the same value when simplified are called **equivalent ratios**. For example, the ratio $\dfrac{2}{5}$ is equivalent, or equal, to $\dfrac{4}{10}, \dfrac{6}{15}, \dfrac{8}{20}$, and so on. Similarly, a ratio of $\dfrac{18}{24}$ is equivalent to $\dfrac{9}{12}$ and $\dfrac{3}{4}$. To find equivalent ratios, we just need to multiply or divide both terms by the same number.

EXAMPLE

▶ A fish tank at a hobby store contains 10 neon tetras and 4 lemon tetras. Another tank has 12 fantail guppies and 6 triangle tail guppies. Are the ratio of neon tetras to lemon tetras and the ratio of fantail guppies to triangle tail guppies equivalent ratios?

▶ Write the two ratios as fractions.

Neon tetras/lemon tetras: $\dfrac{10}{4}$

Fantail guppies/triangle tail guppies: $\dfrac{12}{6}$

▶ Find a common denominator by multiplying: $4 \times 6 = 24$.

▶ Multiply the numerator and denominator by the same number to find equivalent fractions.

$$\frac{10 \times 6}{4 \times 6} = \frac{60}{24}$$

$$\frac{12 \times 4}{6 \times 4} = \frac{48}{24}$$

$$\frac{60}{24} > \frac{48}{24}$$

▶ $\dfrac{10}{4}$ and $\dfrac{12}{6}$ are *not* equivalent ratios.

The following word problem involves finding a missing value in two equivalent ratios.

EXAMPLE

▶ Briana is making punch for a party. The recipe calls for 2 cups of apple juice for every 3 cups orange juice. She pours 15 cups of orange juice into the punch bowl. How much apple juice should Briana add?

▶ Write an equation that represents the facts in the problem.

$$\frac{2}{3} = \frac{x}{15}$$

▶ Find equivalent ratios.

$$\frac{2}{3} = \frac{2 \times 5}{3 \times 5} = \frac{10}{15}$$

▶ Briana should add 10 cups of apple juice to the punch.

Comparing Ratios

Suppose we wanted to compare the ratios $\frac{5}{8}$ and $\frac{3}{5}$ to find out which is greater. We already know how to do this since it is the same thing as comparing two fractions with different denominators. Therefore, the solution involves finding a common denominator.

$$\frac{5 \times 5}{8 \times 5} = \frac{25}{40}$$

$$\frac{3 \times 8}{5 \times 8} = \frac{24}{40}$$

Now, it's very easy to compare the two ratios. In fact, they are very close in value. Nonetheless, since $\frac{25}{40}$ is greater than $\frac{24}{40}$, we can determine that $\frac{5}{8}$ is greater than $\frac{3}{5}$.

Another way to compare the value of two ratios is to find their decimal equivalents. We can find the decimal value of a fraction by dividing the numerator by the denominator. Life is much simpler if we have a calculator handy when we use this method!

EXAMPLE

▶ Which is greater, $\frac{6}{7}$ or $\frac{7}{8}$?

▶ Use a calculator:

$$\frac{6}{7} \approx 0.857$$

$$\frac{7}{8} = 0.875$$

▶ Since 0.875 is greater than 0.857, $\frac{7}{8}$ is greater than $\frac{6}{7}$.

Some situations involve comparing two ratios to determine which is greater, as the example below shows.

▶ Jennifer collects old postcards. She has 25 postcards of famous sites in the United States and 15 postcards of famous sites in Europe. Jennifer's best friend Ruthann has 16 postcards of famous American sites and 8 postcards of famous European sites. Whose collection of postcards has a greater ratio of famous American sites to famous European sites?

▶ Write the ratios for each collection as a fraction.

$$\text{Jennifer's collection:} \frac{\text{American sites}}{\text{European sites}} = \frac{25}{15}$$

$$\text{Ruthann's collection:} \frac{\text{American sites}}{\text{European sites}} = \frac{16}{8}$$

▶ Find the decimal value of each fraction.

$$\frac{25}{15} = \frac{5}{3} \approx 1.67$$

$$\frac{16}{8} = \frac{2}{1} = 2.0$$

▶ Since 2.0 is greater than 1.67, Ruthann's collection of postcards has a greater ratio of American sites to European sites than Jennifer's postcard collection.

Rates

A **rate** is a special type of ratio that compares two quantities measured in different units. When we hear that a car is moving at 35 miles per hour, we are really being given a rate: $\dfrac{35\,\text{miles}}{1\,\text{hour}}$. The two quantities being compared are miles and hours. When we see a ratio between two different units of measure with a denominator of 1, it is called the **unit rate**. Other common examples of unit rates are 25 miles per gallon, $10 per pound, and $15 per hour.

 IRL The word *per* comes from Latin and means "for each." Recall that we do not write 1 before a measure of unit. Therefore, an expression such as "35 miles per hour" really means "35 miles for each 1 hour."

EXAMPLE

Debra is on a long-distance bike race. She biked 39 miles in 3 days. What is Debra's unit rate?

Write the original rate given in the problem.

$$\frac{\text{Miles biked}}{\text{Number of days}} = \frac{39}{3}$$

Simplify the rate by using division to make the denominator 1.

$$\frac{\text{Miles biked}}{\text{Number of days}} = \frac{39 \div 3}{3 \div 3} = \frac{13}{1}$$

Debra biked 13 miles per day.

Some problems ask us to compare two rates to determine which is greater or smaller.

▶ Danielle and Joachim work at the same bakery. Danielle bakes 420 cookies in a 4-hour shift. During his 3-hour shift Joachim bakes 330 cookies. Who baked more cookies per hour, Danielle or Joachim?

▶ Write the rates as stated in the problem and simplify to unit rates with a denominator of 1.

$$\text{Danielle:} \quad \frac{420}{4 \, \text{hours}} = \frac{105}{1}$$

$$\text{Joachim:} \quad \frac{330}{3 \, \text{hours}} = \frac{110}{1}$$

▶ $\frac{110}{1}$ is greater than $\frac{105}{1}$. Therefore, Joachim bakes more cookies per hour than Danielle.

Proportions

A **proportion** is an equation that states that two ratios are equivalent. The numbers in a proportion are called **terms**. The simplest way to determine if two ratios form a proportion is to check the cross products of the proportion's terms. If the cross products are equal, then the two ratios form a proportion. Here's how we can check:

$$\frac{4}{5} = \frac{16}{20}$$

since 4 times 20 equals 80 and 5 times 16 equals 80.

We can find an unknown number in a proportion by solving the proportion. Sometimes, this can easily be done using mental math. For example, when solving $\frac{65}{x} = \frac{13}{6}$, notice that 65 is five times 13. So, the unknown denominator will be five times the known numerator: $5 \times 6 = 30$.

So, x equals 30: $\dfrac{65}{30} = \dfrac{13}{6}$.

We can use cross multiplication to check if $\dfrac{65}{30}$ equals $\dfrac{13}{6}$.

$$65 \times 6 = 13 \times 30$$
$$390 = 390$$

Therefore, $\dfrac{65}{30} = \dfrac{13}{6}$ forms a proportion.

Another way to find the value of the unknown term in a proportion is to use some basic algebra. Here's how we can go about using this method.

EXAMPLE

▶ What is the value of x in the proportion $\dfrac{9}{12} = \dfrac{x}{72}$?

▶ First, write the original proportion.

$$\frac{9}{12} = \frac{x}{72}$$

▶ Next, multiply each side by the denominator of the ratio with the unknown term (x).

$$\frac{9}{12} \times \frac{72}{1} = \frac{x}{72} \times \frac{72}{1}$$

$$\frac{648}{12} = \frac{72x}{72}$$

▶ Finally, solve for x.

$$\frac{648}{12} = \frac{72x}{72}$$

$$\frac{648 \div 12}{12 \div 12} = x$$

$$x = 54$$

▶ Therefore, x equals 54 and $\dfrac{9}{12}$ equals $\dfrac{54}{72}$.

Here's a real-world problem that asks us to compare unit rates.

EXAMPLE

> A supermarket has apple juice on sale. A 32-ounce quart of apple juice costs $3.00 and a 128-ounce gallon of apple juice costs $9.60. Which size container of orange juice offers the better unit price per ounce?

> First, write the price and ounces per container as a ratio.

$$\text{Quart: } \frac{\text{cost}}{\text{ounces}} = \frac{\$3.00}{32} \approx 0.0937 \approx \$0.094 \text{ per fluid ounce}$$

$$\text{Gallon: } \frac{\text{cost}}{\text{ounces}} = \frac{\$9.60}{128} = 0.075 = \$0.075 \text{ per fluid ounce}$$

> $0.075 is less than $0.094. Therefore, a gallon offers the better unit price for orange juice.

Percents

The word *percent* means "out of 100" or "in one hundred." A **percent**, therefore, is a ratio that compares a number to 100. The symbol % stands for percent.

As we've seen earlier, a ratio can be represented in several different ways. For example, the fraction $\frac{3}{5}$ can be represented by the decimal 0.6, and it can also be represented by the equivalent fraction $\frac{60}{100}$. Thus, $\frac{3}{5}$, 0.6, and $\frac{60}{100}$ all express the same value.

Recall that we can find the decimal representation of any fraction simply by dividing the numerator by the denominator. For example, $\frac{6}{8}$ equals 0.75 and $\frac{7}{9}$ equals $0.\overline{77}$. The bar over the 7s indicates that these numbers repeat indefinitely, so $\dfrac{7}{9}$ equals 0.77777777... and so on. Decimals are written as percents simply by dropping the decimal point and adding a percent symbol. So, 0.75 becomes 75% and 0.77777777 rounds to 78%.

Fractions, decimals, percents, and their equivalences can be readily modeled using hundred grids. For example, $\dfrac{43}{100}$, 0.43, and 43% can all be modeled as:

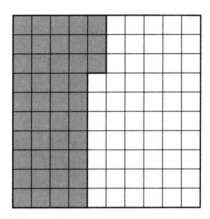

EXAMPLE

▶ How do we write the shaded portion as a fraction, decimal, and percent?

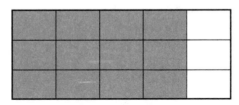

▶ Write a fraction that represents the diagram.

$$\dfrac{\text{Shaded boxes}}{\text{All boxes}} = \dfrac{12}{15}$$

▶ Then, write a decimal that represents the fraction.

$$\frac{12}{15} = 0.80$$

▶ Finally, write a percent that represents the decimal.

$$\frac{12}{15} = 0.80 = 80\%$$

▶ The shaded portion of the diagram can be written as $\frac{12}{15}$, 0.80, or 80%.

There are a number of different types of real-world situations that involve percents. Let's take a look.

Finding the Percent of a Number

The most common type of percent problem is to find the percent of a number. For example, suppose a store sells two brands of cell phones. Of the past 50 sales, the store sold 60% of brand A. How many cell phones of brand A did the store sell?

Recall that every percent can be written as a decimal or as a fraction, so there are two ways to solve the problem:

$$\begin{array}{r} 50 \\ \times\ 0.6 \\ \hline 30.0 \end{array}$$

or

$$\frac{50}{1} \times \frac{60}{100} = \frac{3{,}000}{100} = 30$$

Whether using a decimal or fractions, we can see that the store sold 30 brand A phones.

Most students find multiplying by a decimal an easier method than dealing with fractions since it involves fewer steps. Nevertheless, either method is correct, so it's really your choice how you go about solving this type of problem unless, of course, you're told which method to use!

EXAMPLE

▶ What is 25% of 160?

▶ Multiply 160 by 0.25.

$$
\begin{array}{r}
160 \\
\times\,0.25 \\
\hline
800 \\
320 \\
\hline
4000
\end{array}
$$

▶ Place the decimal point two places to the left.

$$4000 \rightarrow 40.00$$

▶ 25% of 160 is 40.

Here's another example of how to find the percent of a number.

EXAMPLE

▶ What is 15% of 120?

▶ Multiply 120 by 0.15.

$$
\begin{array}{r}
120 \\
\times\ \,0.15 \\
\hline
600 \\
120 \\
\hline
18.00
\end{array}
$$

▶ 15% of 120 is 18.

Let's look at an example that involves finding the percent of a number.

EXAMPLE

Francine bought a new clarinet. Its original price was $800, but the instrument was on sale for 25% off. How much did Francine save on the purchase price of the clarinet?

First multiply the sales price by the sales percent. Recall that 25% equals 0.25.

$$800 \times 25\% = ?$$

$$\$800 \times 0.25 = \$200$$

Francine saved $200 on the purchase price of the clarinet.

BTW

Remember that multiplication is commutative, so it doesn't matter if you multiply 5 by 10 or 10 by 5—the product 50 is still the same. You can use this fact to help you solve percent problems. For example, suppose you want to find 80% of 50. Rather than do the original calculation, simply flip the problem and find 50% of 80. Since 50% is equivalent to $\frac{1}{2}$, all you have to do is think half of 80 is 40. So 80% of 50 is 40!

Sometimes, the percent we are asked to find is greater than 100%.

EXAMPLE

What is 150% of 200?

Write 150% as 1.5, then multiply 200 by 1.5.

```
    200
 ×  1.5
   1000
   200
  300.0
```

or

$$\frac{200}{1} \times \frac{150}{100} = \frac{30,000}{100} = 300$$

▶ 150% of 200 is 300.

Finding What Percent One Number Is of Another Number

Some percent problems ask us to find what percent one number is of another. For example, what percent of 250 is 80? To solve this problem, we must write a percent equation.

$250 \times x\% = 80$	Write an equation.
$x = \dfrac{80}{250}$	Divide both sides by 250.
$= 0.32$	Simplify
$x = 32\%$	Change the decimal to a percent.

A common percent problem involves finding the sales tax rate.

EXAMPLE

▶ Jorge bought a new telescope for $700. He paid $56 in sales tax. What was the tax rate?

▶ Let x equal the tax rate. Write an equation or proportion.

$$700 \times x\% = 56$$
$$x = \frac{56}{700}$$
$$x = 0.08$$
$$x = 8\%$$

▶ Jorge paid a sales tax rate of 8%.

Other problems ask us to find what percent of the original price a sales price is.

▶ Brenda purchased a set of golf clubs for $375. The set was originally priced at $500. What percent of the original price was the sale price?

▶ Let x equal the sales rate. Write an equation or proportion.

$$500 \times x\% = 375$$
$$x = \frac{375}{500}$$
$$x = 0.75$$
$$x = 75\%$$

▶ The sale price was 75% of the original price.

Finding a Number When the Percent Is Known

In some cases, a percent problem asks us to find a number when a percent of the number is known. For example, a footwear manufacturer knows that it sold about 75,000 pairs of its most popular running shoes. These sales account for about 60% of all styles of running shoes sold by the company. About how many pairs of running shoes did the footwear manufacturer sell in all?

To solve this problem, we must write an equation and solve. Think: 60% of what number is 75,000? Write 60% as 0.6:

$$x = 75,000 \div 0.6$$
$$x = 125,000$$

Therefore, the manufacturer sold 125,000 pairs of running shoes in all.
Let's work out a similar problem together.

EXAMPLE

▶ If 80% of a number is 2,000, what is the number?

▶ Write an equation and solve.

$$0.8 \times x = 2,000$$

$$x = \frac{2,000}{0.8}$$

$$x = 2,500$$

▶ 2,000 is 80% of 2,500.

Let's look at a word problem.

EXAMPLE

▶ Mark and Jessica left $10.00 as a tip on a dinner bill at a restaurant. If the tip was 20% of the total bill, how much did Mark and Jessica pay for dinner?

▶ Think of 20% as 0.2. Write an equation.

$$0.2 \times x = 10.00$$
$$0.2x = 10.00$$
$$\frac{0.2x}{0.2} = \frac{10.00}{0.2}$$
$$x = 50$$

▶ Mark and Jessica's dinner bill was $50.00.

Percent of Increase and Percent of Decrease

If we want to show how much a quantity has increased or decreased when compared to the original amount, we must find the **percent of change**. When the new amount is greater than the original amount, the **percent of increase** tells the percent change.

▶ A local biking store sold $22,300 worth of bikes and biking equipment last month. This month the store's total sales are $25,200. What is the percent of increase in sales?

▶ Subtract last month's sales from this month's sales: $25,200 − $22,300 = $2,900.

▶ Write a ratio and find its decimal equivalent.

$$\frac{\text{amount of increase}}{\text{original number}} = \frac{2,900}{22,300} \approx 0.1300$$

▶ Write the decimal as a percent by multiplying times 100: 0.1300 × 100 = 13%.

▶ The percent of increase in sales is approximately 13%.

This next example shows how to calculate the **percent of decrease**.

▶ In May, a man's suit sold for $400. The price was reduced in August to $250. What was the percent of decrease in the price of this suit?

▶ Find the dollar amount of the decrease by subtracting: 400 − 250 = 150.

▶ Write a percent of change equation.

$$\frac{\text{amount of decrease}}{\text{original number}} = \frac{150}{400} = 0.375$$

▶ Write the decimal as a percent by multiplying times 100: $0.375 \times 100 = 37.5\%$.

▶ The percent of decrease in the price of the suit is 37.5%.

Simple Interest

If we borrow money from a bank for a period of **time** (t), we pay an annual **rate of interest** (r) based on the **principal** (P), or the amount we borrowed. If we deposit money in a savings account at the bank, we will receive interest based on a rate and the amount of time we leave the money in the bank.

EXAMPLE

▶ You deposit $2,000 in a savings account at an annual simple rate of interest 3%. You leave the money in the account for three years. How much interest do you earn?

▶ To find the simple interest on the account, use the formula:

$$\text{Interest } (I) = \text{principal } (p) \times \text{rate } (r) \times \text{time } (t)$$

▶ Plug the numbers into the formula and solve.

$$I = 2,000 \times 0.03 \times 3$$
$$= 180$$

▶ In three years, you will earn $180 in interest on a savings account of $2,000 at a simple annual interest rate of 3%.

Here's an example that shows how to calculate the interest on a loan from a bank.

▶ You borrow $20,000 to buy a boat. The bank offers you a 5-year-loan at 6% simple interest. How much interest will you pay on the loan at the end of the five years?

▶ Plug the numbers into the simple interest formula and solve.

$$I = p \times r \times t$$
$$= 20,000 \times 0.06 \times 5$$
$$= 6,000$$

▶ In five years, you will have paid $6,000 in interest on a loan of $20,000 at a simple annual interest rate of 6%.

EXERCISES

EXERCISE 5–1

Solve each problem by writing the ratio.

1. Garret collects old comic books. He owns 5 Superman comics and 9 Batman comics. Write the ratio of Superman comics to Batman comics in three different ways.

2. There are 98 students in the eighth grade at Jefferson Middle School. 52 students are girls and 46 are boys. What is the ratio of boys to girls? Write the ratio in simplest form.

3. A major league baseball team won 96 of the 162 games it played last season. What is the ratio of wins to games played? Write the ratio in simplest form.

4. A TV documentary on Dr. Martin Luther King was an hour long. The hour consisted of 18 minutes of ads and 42 minutes of documentary footage. What was the ratio of advertising minutes to documentary minutes? Write the ratio in simplest form.

EXERCISE 5–2

Solve each problem by finding equivalent ratios or by comparing ratios.

1. What are two equivalent ratios to $3:5$?

2. Mr. Ralston's eighth-grade homeroom class has 36 students. 16 are boys and 20 are girls. The 8th grade at the school has a total of 108 students, 48 boys and 60 girls. Is the ratio of boys to girls in Mr. Ralston's homeroom class the same as the ratio of boys to girls in the entire 8th grade? Explain.

EXERCISE 5–3

Complete the following comparisons by writing >, <, or =.

1. $4:3$ _?_ $32:24$

2. $7:35$ _?_ $21:84$

3. $13:66$ _?_ $11:77$

4. $8:15$ _?_ $24:45$

EXERCISE 5–4

Solve each ratio, rate, or proportion problem.

1. Raul mixes 2 ounces of red paint to 3 ounces of yellow paint to make 5 ounces of orange paint. If Raul wants to make 20 ounces of orange paint, how much red paint and yellow paint should he mix?

2. Bruce types 185 words in 5 minutes. What is Bruce's typing rate of words per minute?

3. Bret has 6 red marbles and 12 blue marbles. Elena has 9 red marbles and 15 blue marbles. Who has the greater ratio of red to blue marbles?

4. Jada's car uses 18 gallons of gasoline to travel 576 miles along a highway. What is her car's rate of miles per gallon?

EXERCISE 5–5

Solve each proportion for the unknown term.

1. What is the value of x in the proportion $\dfrac{9}{x} = \dfrac{3}{4}$?

2. What is the value of x in the proportion $18:12 = 60:x$?

3. A car travels 195 miles in 3 hours. Traveling at the same speed, how far will the car travel in 5 hours? Write the proportion and solve.

4. A 24-pack of 20-ounce orange juice bottles costs $26. How much should a 36-pack of the same juice cost if the price per bottle stays the same?

EXERCISE 5–6

For questions 1 and 2, write the equivalencies described. In questions 3 and 4, solve the percent problems.

1. How do you write $\dfrac{17}{25}$ as a decimal and percent?

2. How do you write 35% as a fraction and decimal?

3. What is 6% of 1,500?

4. What percent of 80 is 4?

EXERCISE 5–7

Solve each word problem involving percent of increase, percent of decrease, or interest.

1. Last year, the school library fundraising event sold 130 raffle tickets. This year, 169 raffle tickets were sold for the fundraiser. What was the percent of increase of raffle tickets sold?

2. Attendance at the art museum last June was 13,884. This June attendance was 11,570 people. What was the percent of decrease in attendance?

3. You buy a $5,000 bond that provides a 5% annual interest rate. How much interest will you have earned after four years?

4. You take out a $10,000, 3-year loan at 6% annual interest to buy a car. How much interest will you have paid after three years?

Flashcard App

Plane Geometry

MUST KNOW

 Plane geometry is the study of two-dimensional figures. Every part of a plane figure lies on the same plane, or flat surface.

 A line is a straight, one-dimensional figure extending forever in opposite directions. A line segment is part of a line that connects two points. A ray is a part of a line with a fixed starting point and extending forever in a particular direction.

 Angles are formed by two rays that meet at a common vertex.

 Triangles are closed plane figures with three straight sides and three angles whose measures add up to 180°.

 The Pythagorean theorem states that the square of the length of the hypotenuse of a right triangle equals the sum of the squares of the lengths of the other two sides: $a^2 + b^2 = c^2$.

 Quadrilaterals are closed plane figures with four straight sides and four angles whose measures add up to 360°.

135

he word *geometry* has its roots in the Greek word *geometrein*, meaning "earth measuring." Geometry itself existed thousands of years before the rise of ancient civilizations in Babylonia, Greece, and India. Measuring land was not the only issue that required ancient cultures to use geometric ideas. Building walls involved ideas such as vertical, parallel, and perpendicular. The usefulness of geometry in early cultures is evident in the fact that its ideas spread rapidly around the world as commerce developed.

Points, Lines, Rays, and Line Segments

Many of the basic terms of geometry have very specific meanings that must be mastered in order to avoid confusion with their meanings in ordinary language. For example, the word *point* often refers to a dot made on a piece of paper. In geometry, a **point** has no size or dimension; instead, it is simply a location in space. By convention, a point is named with a capital letter in italic type, such as "point *A*" or "point *B*." When you draw a line, it will have a beginning and an end. In geometry, however, a **line** is a collection of points that extends forever in opposite directions. Two points are used to name a line. For example, $\overleftrightarrow{XY}$ names this line:

$$X \qquad Y$$

Notice that the line above the letters $\overleftrightarrow{XY}$ has arrows at both ends that point in opposite directions.

Unlike a line, a **ray** starts at one endpoint and extends forever in a particular direction:

$$A \qquad B$$

The ray $\overrightarrow{AB}$ has an arrow pointing off only in the direction where it extends indefinitely. The point where the ray starts has a dot to signify that it is a beginning point.

A **line segment** is part of a line consisting of two endpoints and all the points between them:

C   D

Notice that the line segment $\overline{CD}$ has dots at both ends.

▶ Using the correct symbol, write the name of the figure below.

E   D

▶ The figure shows a ray because it starts at one endpoint and extends forever in a particular direction.

▶ The correct way to write this ray is $\overrightarrow{DE}$.

Let's look at another example.

▶ Write the name of the figure below using the correct symbol.

J   K

▶ The figure shows a line because it extends forever in both directions.

▶ The correct way to write this line is $\overleftrightarrow{JK}$.

BTW

To determine whether you are looking at a line, ray, or line segment, check both ends of the figure for dots and arrows.

The orientation of a ray, line segment, or line does not change the way it is classified.

▶ Write the name of the figure below using the correct symbol.

X

Y

▶ Even though the figure is vertical, it shows a line that extends forever in both directions.

▶ The correct way to write this line is $\overleftrightarrow{XY}$.

Types of Angles

An **angle** is formed when two rays come together at an endpoint. The common endpoint where the two rays meet is called the **vertex**. The rays are referred to as the **sides** of the angle. The symbol $\angle$ is placed before the three capital letters that name the angle. The three letters chosen for the name of an angle are always in alphabetical order, with the middle letter representing the vertex of the angle. The angle below is named $\angle DEF$:

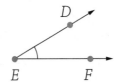

D

E   F

Sometimes, an angle is named by a number:

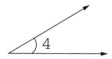

One way to categorize angles is by the measure of degrees (°) at its vertex:

- A **right angle** has sides that are perpendicular (⊥) to each other and has a measure of exactly 90°. By convention, a right angle shows a little square at the vertex to indicate that its measure is 90°.

- An angle that has a measure of less than 90° is called an **acute angle**.

- An **obtuse angle** has a measure greater than 90° but less than 180°.

- A **straight angle** is a line that has a measure of exactly 180°.

Looking at the vertex is the best way to determine the type of angle we are examining.

▶ Is ∠ABC a right, acute, obtuse, or straight angle?

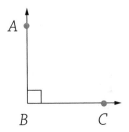

▶ ∠ABC is a *right* angle because the measure of its vertex is exactly 90°. The square drawn in the angle tells us the angle is 90°.

Remember that the measure of the vertex determines what type of angle is represented.

▶ Is ∠DEF a right, acute, obtuse, or straight angle?

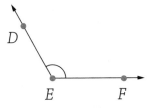

▶ ∠DEF is an *obtuse* angle because the measure of its vertex is greater than 90°.

Now, let's try identifying another common type of angle.

EXAMPLE

▶ Is ∠JKL a right, acute, obtuse, or straight angle?

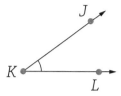

▶ ∠JKL is an *acute* angle because its vertex is less than 90°.

Let's try one more example.

EXAMPLE

▶ Is ∠PQR a right, acute, obtuse, or straight angle?

▶ ∠PQR is a *straight* angle because the measure of vertex is 180°.

Complementary and Supplementary Angles

Two angles are **complementary** if the sum of the measures of their vertices is 90°, that is, when taken together they form a right angle. A quick glance at ∠WXZ and ∠ZXY shows that they form complementary angles:

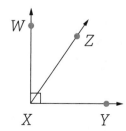

When the sum of the measures of two angles is 180°, they are called **supplementary** angles. In the example below, we can tell that $\angle PQR$ and $\angle RQS$ are supplementary angles because together they form a straight line:

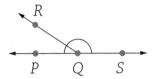

Knowing the measure of a complementary or supplementary angle can help us determine the measure of the other angle in the pair. To find the unknown angle measure, all we need to do is subtract the known measure from the total measure. Consider this pair of angles:

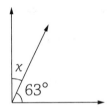

If we are told that the pair of angles form a right angle, we know that the sum of the measures of complementary angles is 90°: $\angle x = 90° - 63° = 27°$.

Now consider this straight line consisting of two angles:

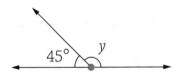

The angles form a straight line, so the sum of the measures is 180°: $\angle y = 180° - 45° = 135°$.

EXAMPLE

What is the measure of $\angle x$?

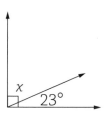

Since this is a right angle, subtract the known measure from 90°.

$$90° - 23° = \angle x$$
$$\angle x = 67°$$

Now, let's try finding the measure of an unknown angle when dealing with supplementary angles.

EXAMPLE

What is the measure of $\angle y$ below?

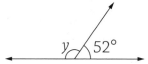

Since the angles form a straight line, they are supplementary angles. Therefore, subtract the known measure from 180°.

$$180° - 52° = \angle y$$
$$\angle y = 128°$$

Angles of Intersecting Lines

When two lines intersect at a point, they create two pairs of opposite angles called **vertical angles**. The opposite angles have the same measure:

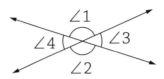

In this figure, $\angle 1$ and $\angle 2$ are the same measure and $\angle 3$ and $\angle 4$ are the same measure. Notice that, when taken together, $\angle 1$ and $\angle 3$ form a straight line, as do $\angle 2$ and $\angle 4$. We can use these relationships to find missing angle measures when dealing with intersecting lines.

EXAMPLE

▶ What are the measures of $\angle B$, $\angle C$, and $\angle D$?

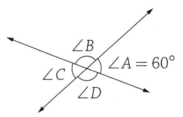

▶ Identify the known angle. The measure of $\angle A = 60°$.

▶ Since $\angle A$ and $\angle C$ are vertical angles, they have the same measure. Therefore, the measure of $\angle C = 60°$.

▶ $\angle A$ and $\angle B$ form supplementary angles. Therefore, the sum of their measures equals $180°$. Use this fact to calculate the measure of $\angle B$.

$$\angle B = 180° - 60° = 120°$$

▶ Since $\angle B$ and $\angle D$ have the same measure, $\angle D$ equals $120°$.

▶ Therefore, $\angle B$ equals $120°$, $\angle C$ equals $60°$, and $\angle D$ equals $120°$.

When two lines are crossed by a third line called a **transversal**, they form corresponding angles. **Corresponding angles** are angles that appear at the same location at each intersection. When a transversal intersects two parallel lines, the corresponding angles have the same measure. In the diagram that follows, the measures of ∠1 and ∠5 are equal. In addition, *m*∠2 equals *m*∠6, *m*∠3 equals *m*∠7, and *m*∠4 equals *m*∠8.

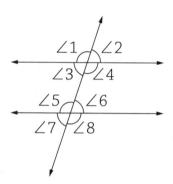

In addition to corresponding angles, when a transversal intersects parallel lines, pairs of alternate interior angles of equal measure are also created. **Alternate interior angles** are two nonadjacent interior angles on opposite sides of a transversal. In the preceding diagram, ∠3 and ∠6 are alternate interior angles, and they have equal measures, as do ∠4 and ∠5. Likewise, **alternate exterior angles** are two nonadjacent exterior angles on opposite sides of a transversal, and these pairs are also equal in measure. Therefore, the measure of ∠1 equals the measure of ∠8 and the measure of ∠2 equals the measure of ∠7.

▶ If the measure of ∠8 equals 75°, what are the measures of ∠1, ∠2, ∠3, ∠4, ∠5, ∠6, and ∠7?

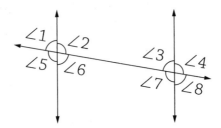

▶ Identify the measure of the known angle, ∠8, which is 75°.

▶ Since ∠8 and ∠3 are vertical angles, they have the same measure. Therefore, the measure of ∠3 is 75°.

▶ ∠7 and ∠8 form supplementary angles that equal 180°. Use this fact to calculate the measure of ∠7: $180° - 75° = 105°$.

▶ Since ∠7 and ∠4 are vertical angles with the same measure, ∠4 equals 105°. You know that ∠3 equals 75°, ∠4 equals 105°, ∠7 equals 105°, and ∠8 equals 75°.

▶ Use what you know about the measures of ∠3, ∠4, ∠7, and ∠8 to find the measures of the corresponding angles ∠1, ∠2, ∠5, and ∠6.

▶ ∠1 equals 75°, ∠2 equals 105°, ∠3 equals 75°, ∠4 equals 105°, ∠5 equals 105°, ∠6 equals 75°, and ∠7 equals 105°.

In geometry, a **plane** is thought of as a flat surface that extends infinitely in all directions. **Plane figures** are flat two-dimensional shapes that can be made of straight lines, curved lines, or a combination of both. A polygon is a closed plane figure made of three or more straight lines and angles. All the examples shown on the next page are plane figures, but those with curved lines are not considered polygons:

Triangles and Quadrilaterals

A **triangle** is a closed plane figure, or polygon, with three straight sides and three angles. The sum of the angles of a triangle equals 180°. We name triangles by the three capital letters used to label its vertices. A triangle is written with the triangle symbol before the three vertices—for example, $\triangle ABC$:

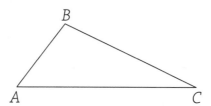

Triangles are classified into three basic groups, by the measure of their angles or by the length of their sides. If we classify triangles according to their angles:

- An **acute triangle** contains three acute angles.

- A **right triangle** contains one right angle.

- An **obtuse triangle** has one obtuse angle.

Different terms are used to categorize triangles by the lengths of their sides:

- In an **equilateral triangle**, all three sides are the same length. In addition, all three angles of an equilateral triangle measure 60°, so this type of triangle is also *equiangular*.

- An **isosceles triangle** has two sides that are the same length.

- In a **scalene triangle**, all three sides have different lengths.

We can summarize the key points to remember about each type of triangle:

acute	3 angles each less than 90°	
right	1 angle that equals 90°	
obtuse	1 angle greater than 90°	
equilateral	3 sides that are the same length	
isosceles	2 sides that are the same length	
scalene	3 sides all different lengths	

Now, let's try identifying a triangle based on the measure of its angles.

EXAMPLE

▶ Based on its angle measures, what type of triangle is △*JKL*? How do you know?

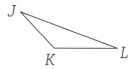

▶ Look at each angle. Notice that the measure of ∠*JKL* is greater than 90°. Therefore, this figure is an obtuse triangle.

Notice how small tick marks are used to indicate sides of the same length in the following example.

EXAMPLE

▶ What type of triangle is △*GHI*? How do you know?

▶ Look at the lengths of the sides. Notice that two of the sides ($\overline{GH}$ and $\overline{HI}$) are the same length.

▶ △*GHI* is an isosceles triangle because two of its sides are the same length.

The **Pythagorean theorem** states that the square of the **hypotenuse** (the side opposite the right angle) of a right triangle is equal to the sum of the squares of the other two legs of the triangle:

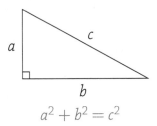

$$a^2 + b^2 = c^2$$

If we know the length of two sides of a right triangle, we can find the length of the third side by using this theorem.

EXAMPLE

▶ What is the length of side b in the triangle shown below?

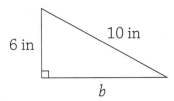

▶ Since this is a right triangle, use the Pythagorean theorem to find the length of missing side: $a^2 + b^2 = c^2$.

▶ Substitute the values you know in the formula. The vertical side a equals 6 in and the hypotenuse side c equals 10 in.

$$6^2 + b^2 = 10^2$$
$$36 + b^2 = 100$$
$$36 - 36 + b^2 = 100 - 36$$
$$b^2 = 64$$
$$b = \sqrt{64}$$
$$b = 8$$

▶ So, the length of side b is 8 in.

There are two special types of right triangles. The first is a **45°-45°-90°** right triangle. Suppose the length of each leg of this triangle is 1 foot. What is the length of its hypotenuse?

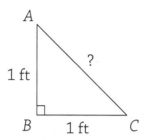

As with all right triangles, we can use the Pythagorean theorem to find the hypotenuse of this triangle.

$$a^2 + b^2 = c^2$$
$$1^2 + 1^2 = c^2$$
$$1 + 1 = c^2$$
$$2 = c^2$$
$$c = \sqrt{2}$$

Thus, the length of the hypotenuse of any 45°–45°–90° is the length of one of its legs times $\sqrt{2}$.

Let's work out a problem together.

▶ What is the length of the hypotenuse of the triangle shown below?

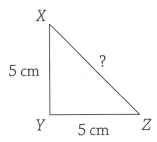

▶ Since the measure each leg of this right triangle is 5 cm, we know it is a 45°–45°–90° triangle.

▶ Therefore the length of hypotenuse XY is $5\sqrt{2}$ cm.

If we only know the length of the hypotenuse of a 45°–45°–90° triangle, we can use the same relationships to find the length of each leg.

▶ What is the length of each leg of the triangle shown below?

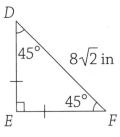

▶ The length of the hypotenuse of this 45°–45°–90° triangle is $8\sqrt{2}$ in.

▶ Therefore, the length of each leg (DE and FE) is 8 in.

The second special right triangle is a **30°-60°-90°** triangle. In any 30°–60°–90° triangle, the hypotenuse is twice the length of the shorter leg. The length of the longer leg is the product of the shorter leg times $\sqrt{3}$.

EXAMPLE

▶ What is the length of leg AB and leg BC in the triangle shown below?

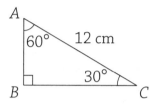

▶ The length of the hypotenuse of this 30°–60°–90° triangle is 12 cm.

▶ First, find the length of the shorter leg by dividing the hypotenuse by 2: $12 \div 2 = 6$. Therefore, the length of AB is 6 cm.

▶ Next, find the length of the longer leg by multiplying the length of the shorter leg times $\sqrt{3}$. Therefore, the length of BC is $6 \times \sqrt{3}$, or $6\sqrt{3}$.

▶ The length of AB is 6 cm and the length of BC is $6\sqrt{3}$ cm.

A **quadrilateral** is a polygon that has four straight sides and four vertices. The sum of the angles of a quadrilateral always equals 360°. A quadrilateral is named according to the letters used to label its vertices—for example, $PQRS$:

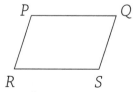

There are several common quadrilaterals that have special features. Take time now to become familiar with their names and key characteristics:

parallelogram	2 pairs of parallel sides	
rectangle	a parallelogram with 4 right angles	
square	a parallelogram with 4 right angles and 4 sides of equal length	
rhombus	a parallelogram with 4 sides of equal length	
trapezoid	exactly one pair of parallel lines	

EXAMPLE

▶ What type of quadrilateral is *ABCD*? How do you know?

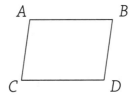

▶ Look at each side. Notice that there are two pairs of opposite, equal, and parallel sides.

▶ *ABCD* is a parallelogram because it has two pairs of opposite, equal, parallel sides.

Now, let's identify the quadrilateral in the following example based on its sides.

▶ What type of quadrilateral is *JKLM*? How do you know?

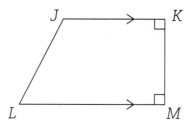

▶ Look at the opposite sides and angles. Notice that only one pair of sides is parallel.

▶ *JKLM* is a trapezoid because it has one pair of parallel sides.

Other Common Polygons

As we noted earlier, a polygon is a closed plane figure that is formed by joining three or more line segments at their endpoints to form vertices. We've already studied two types of polygons—triangles comprised of three line segments and quadrilaterals with four line segments.

You might be surprised to find out that there are polygons with many, many more sides. For example, a hectogon is 100-sided polygon, a chiliagon is 1,000-sided polygon, and a megagon is 1,000,000-sided polygon. Fortunately, we don't have to study these figures (!), but it's worthwhile to explore the characteristics of polygons with as many as 8 sides.

Recall that the sum of the interior angles of a triangle is 180°, and the sum of the interior angles of a quadrilateral is 360°. A simple formula lets us calculate the sum of the interior angles of any polygon:

$$S = (n - 2)(180°)$$

In this formula, S refers to the sum of the angles, and n equals the number of sides of the polygon. So, all we really have to do is subtract 2 from the number of sides and multiply by 180°. Let's apply the formula to a hexagon, or 6-sided figure:

$$S = (n - 2)(180°)$$
$$= (6 - 2)(180°)$$
$$= (4)(180°)$$
$$= 720°$$

The sum of the interior angles of a hexagon, therefore, equals 720°. This fact holds true if the figure is a regular hexagon or an irregular one:

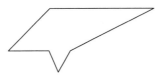

This hexagon may look very strange, but it has 6 sides so the sum of its interior angles is still 720°!

EXAMPLE

▶ What is the sum of the measures of the interior angles of this polygon?

▶ The figure is an octagon with 8 sides. Use the formula $S = (n - 2)(180°)$.

$$S = (8 - 2)(180°)$$
$$= 6(180°)$$
$$= 1,080°$$

▶ The sum of the interior angles of an octagon is 1,080°.

Since all interior angle measures in a *regular* polygon are equal, to find the measure of one of its angles we need only divide by its number of sides. We just saw that the sum of the interior angles of an octagon is 1,080°. Therefore, each angle of a regular octagon: $1{,}080 \div 8 = 135°$. We can take the interior angle formula but divide it by n, the number of sides:

$$m \angle \text{interior angle} = \frac{(n-2)(180°)}{n}$$

This formula enables us to calculate the measure of an interior angle of any a regular polygon.

EXAMPLE

▶ What is the measure of an interior angle of this regular polygon?

▶ The figure is a regular hexagon with 6 sides. Use the formula $m \angle \text{interior angle} = \dfrac{(n-2)(180°)}{n}$.

$$m \angle \text{interior angle} = \frac{(6-2)(180°)}{6}$$
$$= \frac{720°}{6}$$
$$= 120°$$

▶ Each interior angle of a regular hexagon is 120°.

This chart shows the key characteristics of all the polygons we've studied so far.

Polygon	Regular Shape	Number of Sides	Sum of Interior Angles	Interior Angle of a Regular Polygon
triangle		3	180°	60°
quadrilateral		4	360°	90°
pentagon		5	540°	108°
hexagon		6	720°	120°
heptagon		7	900°	≈128.57°
octagon		8	1,080°	135°

Congruent and Similar Figures

Congruent polygons have the same size and shape. In other words, each pair of corresponding sides and corresponding angles of congruent polygons has identical measures. The symbol for congruence is ≅. Based on the figure below, we can write △ABC ≅ △DEF:

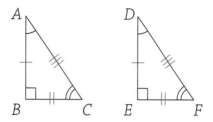

We use tick marks to identify the sides of the triangles that are congruent. Similarly, angle marks are used to indicate angles that have the same measure. In △ABC and △DEF, side AB is congruent to side DE, side BC is congruent to side EF, and side AC is congruent to side DF. Likewise, each angle of △ABC has a corresponding and congruent angle in △DEF.

Both △JKL and △MNO, below, have the same shape and angle measures. However, they are not congruent because their sides are not the same length:

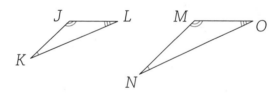

The sides of these two triangles are proportional to each other. In short, one is a scale drawing of the other, but they are not identical. If we tried to place △JKL over △MNO, it would not cover it. **Similar figures** have the same shape and angle measures but are not the same size.

Knowing that two figures are congruent or similar can help us find an unknown angle measure or length.

▶ Quadrilaterals *PQRS* and *TUVW* are congruent. What is the measure of side $\overline{TU}$?

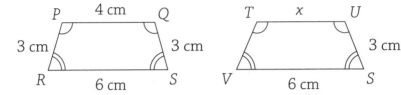

▶ Because $\overline{PQ}$ is congruent with side $\overline{TU}$, the two sides have the same measure. Both $\overline{TU}$ and $\overline{PQ}$ are 4 cm long.

The corresponding angles in pairs of similar figures are always the same measure. However, the measures of their sides are proportional; that is, they share a common ratio.

▶ Quadrilaterals *GHJK* and *LMNO* are similar figures. What is the measure of side $\overline{MN}$?

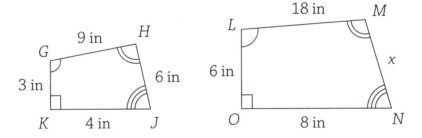

▶ Use the sides for which the measures are known to find the ratio of the corresponding sides of similar figures. For the preceding quadrilaterals, sides $\overline{LO}$ and $\overline{GK}$ and sides $\overline{LM}$ and $\overline{GH}$ are corresponding sides with known measures.

$$\frac{6}{3} = \frac{18}{9} = \frac{2}{1}$$

▶ Since side $\overline{HJ}$ and side $\overline{MN}$ are corresponding sides and the ratio of the lengths of these sides is 2 to 1, side $\overline{MN}$ is twice as long as side $\overline{HJ}$.

$$6 \times 2 = 12$$

▶ Side $\overline{MN}$ is 12 in long.

Perimeter and Area of Polygons

The perimeter of a polygon is the distance around the figure. In other words, the **perimeter** is the sum of the measures of the lengths of the sides of a polygon. To find the perimeter simply add the lengths of all sides.

EXAMPLE

▶ What is the perimeter of $\triangle RST$?

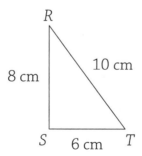

▶ Add the lengths of the three sides of the triangle: $8 + 10 + 6 = 24$.

▶ The perimeter of $\triangle RST$ is 24 cm.

Here's an example of how to find the perimeter of a slightly irregular pentagon.

▶ What is the perimeter of the pentagon *JKLMN*?

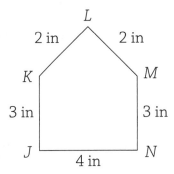

▶ Add the lengths of the five sides of the pentagon.

$$2 \text{ in} + 2 \text{ in} + 3 \text{ in} + 4 \text{ in} + 3 \text{ in} = 14 \text{ in}$$

▶ The perimeter of pentagon *JKLMN* is 14 in.

No matter how irregular a polygon may be, we find the perimeter by adding the measures of its sides.

▶ What is the perimeter of the polygon below?

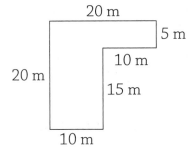

▶ Add the lengths of the six sides of the polygon.

$$20 \text{ m} + 5 \text{ m} + 10 \text{ m} + 15 \text{ m} + 10 \text{ m} + 20 \text{ m} = 80 \text{ m}$$

▶ The perimeter of the polygon is 80 m.

The **area** of a plane figure is the measure of the number of square units needed to cover its surface. If we look at the rectangle below, we can see that it consists of 12 square units:

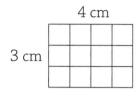

4 cm

3 cm

Each unit is 1 square centimeter (1 cm^2). Therefore, the area of the rectangle is: 3 cm × 4 cm = 12 cm^2. Notice that when we multiply centimeters by centimeters, the product is written as cm^2.

The formula for finding the area of a rectangle is $A = lw$, where A equals area, l equals the length of the figure, and w equals its width. We find the area of a square in much the same way. However, since all sides of a square are equal, it's simpler to express the area formula as $A = s^2$, where s is the length of a side.

EXAMPLE

▶ Raphael is working with his father to make a vegetable garden. The garden is a rectangle that is 10 feet long and 8 feet wide. What is the area of the vegetable garden?

8 ft

10 ft

▶ To find the area of the garden, multiply length by width.

$$A = 10 \text{ ft} \times 8 \text{ ft} = 80 \text{ ft}^2$$

▶ The area of the vegetable garden is 80 ft^2.

Finding the area of a parallelogram is similar to finding the area of a rectangle. As we know, the area of a rectangle is $A = lw$. The area of a parallelogram is the product of its base (b) and its height (h), or $A = bh$. Consider how we can use the height of a parallelogram to create a rectangle.

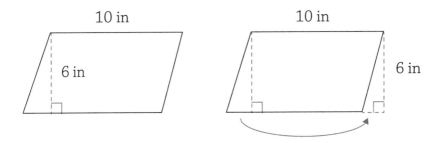

In this case, the area of the parallelogram shown is: $A = 10 \text{ in} \times 6 \text{ in} = 60 \text{ in}^2$.

EXAMPLE

▶ Mason is creating a painting with a canvas in the shape of a parallelogram that is 24 inches long and 18 inches high. What is the area of the canvas?

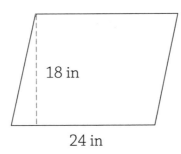

▶ Use the formula for finding the area of a parallelogram: $A = bh$.

$$A = 24 \text{ in} \times 18 \text{ in} = 432 \text{ in}^2$$

▶ The area of the canvas is 432 in^2.

Now that we know how to find the area of quadrilaterals, we'll find it easy to understand and remember the formula for finding the area of a triangle.

The area of a triangle is simply half the area of a rectangle or parallelogram with the same base and height. Therefore, to find the area of a triangle, we can use the formula $A = \dfrac{1}{2}bh$.

EXAMPLE

▶ What is the area of $\triangle ABC$?

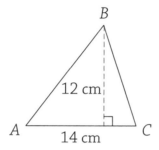

▶ Use the formula for the area of a triangle to find the answer.

$$A = \frac{1}{2}bh$$

$$= \frac{1}{2}(14 \text{ cm} \times 12 \text{ cm})$$

$$= \frac{1}{2}(168 \text{ cm}^2)$$

$$= 84 \text{ cm}^2$$

BTW

Notice that the area of plane figures is always expressed in square units of measurement such as ft², cm², yd², and mi². It's essential to write the exponent next to the unit of measure.

▶ The triangle has an area of 84 cm².

Circumference and Area of a Circle

Polygons have sides that are straight lines, but a circle does not. Instead, a **circle** is the set of all points in a plane that are the same distance from a given point called the **center**. The distance around a circle is called the **circumference** (C).

The **diameter** (*d*) of a circle is a line segment that passes through its center and has both endpoints on the circle. The **radius** (*r*) is a line segment that has one endpoint at the center of a circle and the second endpoint on the circle. The radius of a circle is half the length of the diameter. The diagram below shows the names of the various parts of a circle.

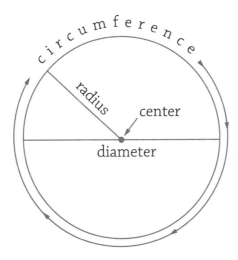

Pi (π) is the ratio of the circumference of a circle to its diameter: $\dfrac{C}{d} = \pi$. It is irrelevant how small or how large the circle is; its circumference is always π times its diameter: $C = \pi d$. Therefore, all we really need to know to find the circumference of a circle is its diameter.

We also can find the circumference of a circle if we know the measure of its radius since the diameter is equal to twice the radius ($d = 2r$). In other words, another way to calculate the circumference of a circle is to use the formula $C = 2\pi r$. Recall that pi is an irrational number (a nonterminating, nonrepeating decimal), so we usually use $\dfrac{22}{7}$ or 3.14 as *approximations* of the value of pi.

EXAMPLE

▶ What is the circumference of the circle below to the nearest tenth, using 3.14 for π?

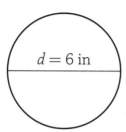

$d = 6$ in

▶ Use the formula for the circumference of a circle to find the answer.

$C = \pi d$
$\quad = 3.14 \times 6$
$\quad \approx 18.84$ in

▶ The circumference of the circle to the nearest tenth is 18.8 in.

Here's an example of how we can find the circumference of a circle using $\dfrac{22}{7}$ for π.

EXAMPLE

▶ What is the circumference of the circle below to the nearest tenth, using $\dfrac{22}{7}$ for π?

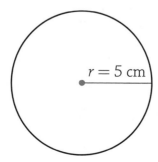

$r = 5$ cm

▶ Use the formula for the circumference of a circle to find the answer.

$$C = 2\pi r$$
$$= 2 \times 5 \times \frac{22}{7}$$
$$= 10 \times \frac{22}{7}$$
$$= \frac{220}{7}$$
$$\approx 31.4 \text{ cm}$$

▶ The circumference of the circle to the nearest tenth is 31.4 cm.

The radius and pi are also used to find the area of a circle, but the formula is different from the one used to find its circumference. It shouldn't come as surprise that the formula has an exponent of 2—after all, the measure of the area of a circle represents two dimensions. The formula for finding the area of a circle is $A = \pi r^2$.

▶ What is the area of the circle below to the nearest tenth, using 3.14 for π?

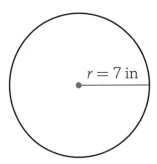

$r = 7$ in

▶ Use the formula for the area of a circle to find the answer.

$$A = \pi r^2$$
$$= \pi(7^2)$$
$$= 49\pi$$
$$= 49 \times 3.14$$
$$\approx 153.9 \text{ in}^2$$

▶ The area of the circle to nearest tenths place is 153.9 in^2.

Here's an example of how to find the area of a circle when we know its diameter.

EXAMPLE

▶ What is the area of the circle below to the nearest hundredth, using 3.14 for π?

▶ Use the formula for the area of a circle to find the answer: $A = \pi r^2$.

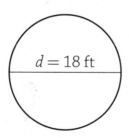

$$d = 18 \text{ ft}$$

▶ Remember that the length of the radius of a circle is half its diameter. Since the diameter is 18 feet, the radius is 9 feet.

$$A = \pi r^2$$
$$= \pi(9^2)$$
$$= 81\pi$$
$$= 81 \times 3.14$$
$$= 254.34$$

▶ The area of the circle to the nearest hundredths place is 254.34 ft^2.

EXERCISES

EXERCISE 6–1

Name each figure using the correct symbol.

1.
    R   S

2.
    P   Q

3.
    L   M

4.
    T   U

EXERCISE 6–2

Identify each angle as right, acute, obtuse, or straight.

1.

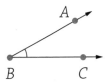

2.

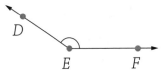

3.

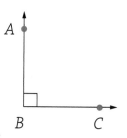

4.

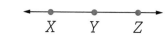

EXERCISE 6-3

Identify each triangle as acute, right, or obtuse.

1.

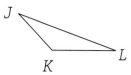

2.

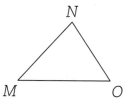

3.

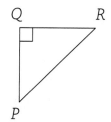

EXERCISE 6–4

Identify each triangle as equilateral, isosceles, or scalene.

1.

2.

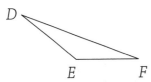

3.

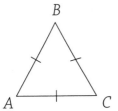

EXERCISE 6–5

What is the length of the unknown side of each triangle?

1.

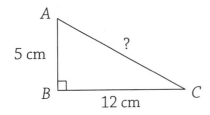

2.

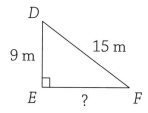

3.

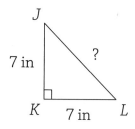

4.

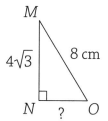

EXERCISE 6–6

What is the measure of ∠A?

1.

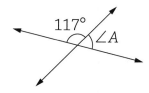

2.

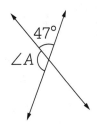

Use the diagram that follows to answer questions 3 and 4.

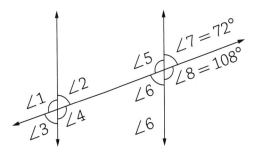

3. What is the measure of $\angle 3$?

4. What is the measure of $\angle 4$?

EXERCISE 6–7

What is the measure of $\angle x$?

1.

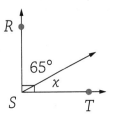

2.

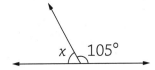

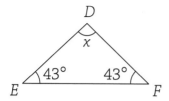

3.

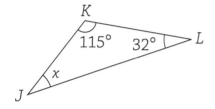

4.

EXERCISE 6–8

Find the sum of the measures of the interior angles for each polygon shown in questions 1 and 2.

1.

2.

Find the measure of each angle of the regular polygons in questions 3 and 4.

3.

4.

EXERCISE 6–9

Find the perimeter of the polygons in questions 1 and 2.

1.

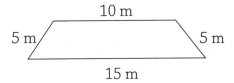

10 m

5 m 5 m

15 m

2.

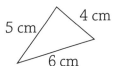

4 cm

5 cm

6 cm

Find the area of the polygons in questions 3 and 4.

3.

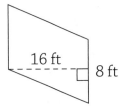

16 ft

8 ft

4.

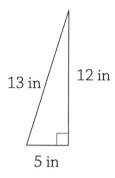

EXERCISE 6–10

Find the circumference of the circles in questions 1 and 2 to the nearest tenths place. Use π = 3.14.

1.

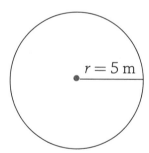

2.

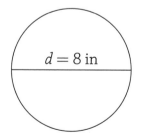

Find the area of the circles in questions 3 and 4 to the nearest tenths place. Use π = 3.14.

3.

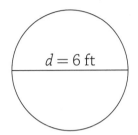

$d = 6$ ft

4.

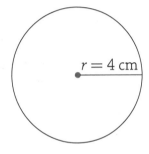

$r = 4$ cm

Flashcard App

 **Solid Geometry**

MUST KNOW

- Solid geometry is the study of three-dimensional objects, such as prisms, pyramids, cylinders, and spheres, that exhibit length, width, and height.

- The faces of a solid figure are polygons that form its sides, its edges are the lines where two faces meet, and its vertices are the points where three edges meet.

- The surface area of a solid figure is the total area of its faces.

- The volume of a solid figure is the amount of space it encloses.

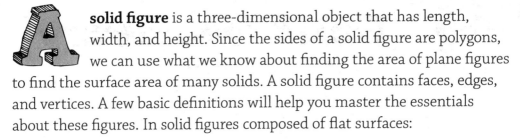

A solid figure is a three-dimensional object that has length, width, and height. Since the sides of a solid figure are polygons, we can use what we know about finding the area of plane figures to find the surface area of many solids. A solid figure contains faces, edges, and vertices. A few basic definitions will help you master the essentials about these figures. In solid figures composed of flat surfaces:

- Each flat surface forms a boundary and is called a **face**.

- An **edge** is a line that is formed where two faces meet.

- A point where three edges meet is called a **vertex**.

Surface Area of Rectangular Prisms

A regular rectangular prism is a solid object that has 6 faces (all of which are rectangles), 12 edges, and 8 vertices. All the opposite sides of a rectangular prism are parallel and congruent.

A cube is really just a special type of rectangular prism. A **cube** is a solid figure with six congruent sides; that is, all the sides have the same dimensions.

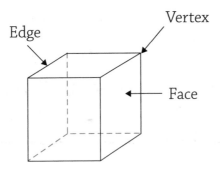

What would be the surface area of a cube with all edges that measure 4 centimeters? A good way to think about how to find the surface area of a cube is to create a **net**, a two-dimensional drawing that can be folded into a three-dimensional solid. The net of a cube looks like:

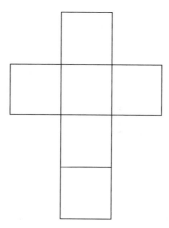

As the net of the cube makes clear, the first thing to do is find the area of one face using the formula $A = s^2$. We can tell that a cube has 6 identical square faces, so the surface area of the entire cube is: $6 \times s^2$. Thus, to find the surface area of a cube we use the formula:

$$SA = 6s^2$$

▶ When fully assembled, a storage box used for moving forms a cube with each edge 30 centimeters in length. What is the surface area of the box to the nearest centimeter?

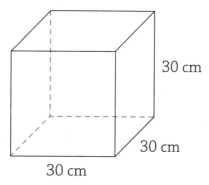

30 cm

30 cm

30 cm

▶ Use the formula for finding the surface area of a cube: $SA = 6s^2$.

$$SA = 6(30 \text{ cm})^2$$
$$= 6(30 \text{ cm} \times 30 \text{ cm})$$
$$= 6(900 \text{ cm}^2)$$
$$= 5{,}400 \text{ cm}^2$$

▶ The surface area of the box is $5{,}400 \text{ cm}^2$.

A **rectangular prism** is a three-dimensional figure that has six faces, all of which are rectangles. Opposite sides of a rectangular prism are parallel to each other and congruent. When all of the angles in a rectangular prism are right angles, the figure is sometimes called a right rectangular prism to distinguish it from an oblique rectangular prism, which is slanted.

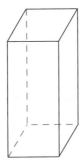 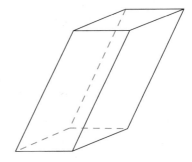

Right rectangular prism   Oblique rectangular prism

The diagram of the rectangular prism on the next page has opposite faces that are congruent: the top and the bottom (length × width) are congruent and opposite sides (length × height) are also congruent. To find the surface area of a rectangular prism, we must find the surface area of each type of face, multiply each measure by 2, and then find the sum of the measures.

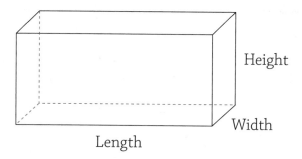

The relationships between these opposite faces make finding the surface area of rectangular prism quite logical, as we can see by the net that appears below.

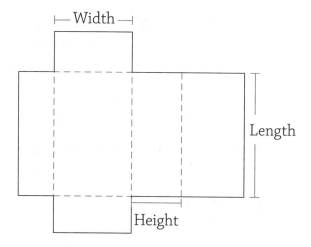

We use the following formula to find the volume of a rectangular prism:

$$SA = 2lw + 2lh + 2wh$$

▶ A jewelry box has a length of 14 inches, a width of 10 inches, and a height of 4 inches. What is the surface area of the box? Use the formula for finding the surface area of a rectangular prism: $SA = 2lw + 2lh + 2wh$.

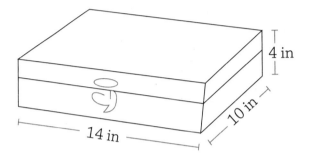

▶ First, find the surface area of either the top or bottom of the jewelry box by multiplying the length by the width and then multiplying the product by 2.

> Top/Bottom: 14 in × 10 in = 140 in^2
> → 140 in^2 × 2 = 280 in^2

▶ Next, find the area of the two long sides by multiplying the length by the height and once again multiplying the product by 2.

> Long Side: 14 in × 4 in = 56 in^2
> → 56 in^2 × 2 = 112 in^2

▶ Then, find the area of the two short sides by multiplying the width by the height and then multiplying by 2.

> Short Side: 10 in × 4 in = 40 in^2
> → 40 in^2 × 2 = 80 in^2

▶ Finally, find the surface area of the box by adding all the measures together.

> $SA = 280$ in^2 + 112 in^2 + 80 in^2
> $= 472$ in^2

▶ The total surface area of the jewelry box is 472 in^2.

Surface Area of Cylinders

Cans of juice, soda, and soup are all examples of cylinders. A **cylinder** is a three-dimensional solid with two circular bases that are opposite and parallel to each other and a curved face that connects the two bases.

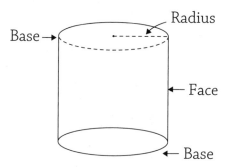

The net that follows shows how the surface area of a cylinder is comprised of three distinct parts, the area of its two circular bases and the area of its face.

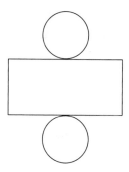

To find the area of the two circles that represent the bases (the top and bottom) of the cylinder, calculate the area of one of the circles (πr^2) and then multiply by 2—in other words, find the value of $2\pi r^2$. Notice that the face of the cylinder forms a rectangle. The width of this rectangle is the same as the height (h) of the cylinder. The length of the rectangle is the same measure as the circumference of one of the circles ($2\pi r$). So, the area of the rectangle is the product of $2\pi rh$. To find the volume of a cylinder, then, we use the formula:

$$SA = 2\pi rh + 2\pi r^2$$

▶ A can of peas has a diameter of 8 centimeters and a height of 11 centimeters. What is the surface area of the can to the nearest tenth?

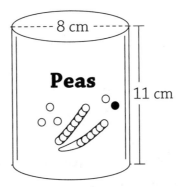

▶ Use the formula for finding the surface area of a cylinder: $SA = 2\pi rh + 2\pi r^2$. Remember that the radius of a circle is half its diameter; therefore, in this example r equals 4 cm.

$$\begin{aligned}
SA &= 2(3.14)(4 \times 11) + 2(3.14)(4^2) \\
&= 2(3.14)(44) + 2(3.14)(16) \\
&= 2(138.16) + 2(50.24) \\
&= 276.32 + 100.48 \\
&= 376.8 \text{ cm}^2
\end{aligned}$$

▶ The surface area of the can is 376.8 cm^2.

Surface Area of Cones

There's a good chance you know what a cone shape looks like from eating ice cream on one. You also may be familiar with traffic cones that are sometimes used when workers are repairing roads and need to redirect traffic.

In geometry, a **cone** is a solid, three-dimensional object with a circular base and a face that tapers smoothly upward to a single point called a **vertex**. The slant height is the length along the face of the cone from its base to its vertex. The radius of the base is the length from the center of the base to a point along its edge. The figures below identify each of the parts of a cone.

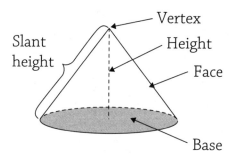

It's important to remember that a cone is a closed figure that has a base. We can see the face and base in the net of the cone that follows.

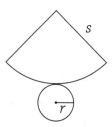

We use the following formula to find the surface area of a cone:

$$SA = \pi rs + \pi r^2$$

Note that, in this formula, πrs represents the area of the face of the cone and πr^2 refers to the area of its circular base.

▶ A peppershaker in the shape of a cone has a radius of 1.5 cm at its base and a slant height of 3 cm. What is the surface area of the peppershaker to the nearest hundredth?

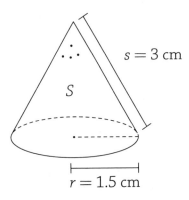

$s = 3$ cm

S

$r = 1.5$ cm

▶ Use the formula for finding the surface area of a cone: $SA = \pi rs + \pi r^2$.

$$SA \text{ (cone)} = (3.14 \times 1.5 \times 3) + (3.14)(1.5^2)$$
$$= (3.14)(4.5) + (3.14)(2.25)$$
$$= 14.13 + 7.065$$
$$= 21.195$$
$$\approx 21.20 \text{ cm}^2$$

▶ The surface area of the peppershaker is approximately 21.20 cm^2.

Surface Area of Pyramids

You probably have an idea about what pyramids look like based on images of the ancient Egyptian pyramids. In geometry, a **pyramid** is a solid figure in which the **base** is a polygon and the faces are triangles that rise up to a common **vertex**. The shape of the base is used to name the pyramid. The diagram that follows shows a square pyramid.

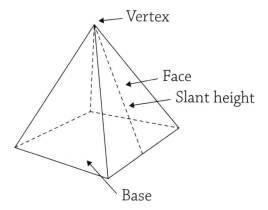

The net of a square pyramid offers a good idea of the components that make up its surface area:

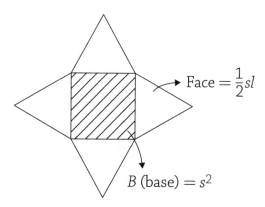

Therefore, to find the surface area of a square pyramid, we must find the area of the square base plus the sum of the area of all its triangular sides. Thus, we use the following formula to find the surface area of a square pyramid:

$$SA = B + 4\left(\frac{1}{2}sl\right)$$

Note that in this formula, B is the area of the base, s is the slant height of a face, and l is its length. The length of the face is the same as the measure of one edge of its base.

▶ A square pyramid has sides that are 4 feet long. Each triangular face has a slant height of 5 feet. What is the surface area of the pyramid to the nearest foot? Use the formula for finding the surface area of a square pyramid: $SA = B + 4\left(\dfrac{1}{2}sl\right)$.

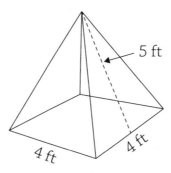

5 ft

4 ft 4 ft

▶ First, find the surface area of the base. Use the formula $A = s^2$.

$$SA = 4\text{ ft} \times 4\text{ ft} = 16\text{ ft}^2$$

▶ Next, find the area of one triangular face. Use the

formula $A = \dfrac{1}{2}sl$. Remember that s is the slant

height of the face and l is the length of one edge of the base.

$$SA = \frac{1}{2}(5\text{ ft} \times 4\text{ ft}) = 10\text{ ft}^2$$

▶ Multiply the surface area of the face by 4 to find the sum of the four face areas.

$$SA = 4 \times 10\text{ ft}^2 = 40\text{ ft}^2$$

▶ Finally, add the areas of the base and the faces.

$$SA = 16\text{ ft}^2 + 40\text{ ft}^2 = 56\text{ ft}^2$$

▶ The surface area of this square pyramid is 56 ft².

BTW

Any polygon may form the base of a pyramid, not just squares. A pyramid with a base formed by a triangle is called a triangular pyramid, a pyramid with a base in the shape of a pentagon is called a pentagonal pyramid, a pyramid with a hexagonal base is called a hexagonal pyramid, and so on.

Surface Area of Spheres

If you've ever held a ball in your hand, you know intuitively what a sphere is. A **sphere** is a three-dimensional figure in the shape of a ball, with every point on its surface the same distance from a fixed point called the **center**.

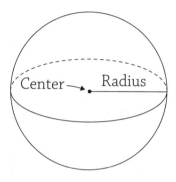

We use the following formula to find the volume of a sphere:

$$SA = 4\pi r^2$$

EXAMPLE

▶ A globe has a diameter of approximately 30 cm. What is the surface area of the globe to the nearest centimeter?

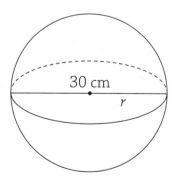

▶ Recall that the radius of a sphere is half its diameter. So, the radius of this sphere is 15 centimeters.

▶ Use the formula for finding the surface area of a sphere: $SA = 4\pi r^2$.

$$SA = 4 \times 3.14 \times 15^2$$
$$= 4 \times 3.14 \times 225$$
$$= 2{,}826 \text{ cm}^2$$

▶ The surface area of the globe is 2,826 cm^2.

Volume of Rectangular Prisms

The **volume** of a solid figure is the amount of space it occupies. Since volume is always a measure of three dimensions, cubic units of measurement are used:

cubic centimeters	cm^3
cubic inches	in^3
cubic feet	ft^3
cubic meters	m^3

In the figure of a rectangular prism that follows, each cube measures 1 cm on each side. How can we find the volume of this rectangular prism?

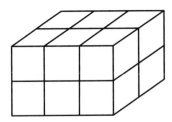

It's easy to count the cubic centimeters to find the volume of this rectangular prism. That might be okay for such a simple example, but matters can get much more complicated: the number of cubes can be much larger and there may not be a grid! A sure-fire way to find the volume of any rectangular prism is to multiply the length (*l*) by the width (*w*) by the height (h). Therefore, we can find the volume of the rectangular prism in this example simply by multiplying: $3 \times 2 \times 2 = 12$ cubic centimeters, or 12 cm^3. Notice that the exponent in the answer is 3, which indicates a cubic measurement. We use the following formula to find the volume of a rectangular prism:

$$V = lwh$$

EXAMPLE

▶ A cube-shaped box measures 12 inches along each edge. What is the volume of the box to the nearest inch?

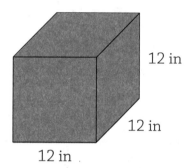

12 in

12 in

12 in

▶ Use the formula for finding the volume of a rectangular prism: $V = lwh$.

$$V = 12 \times 12 \times 12$$
$$= 1,728 \text{ in}^3$$

▶ The volume of the box to the nearest cubic inch is 1,728 in^3.

BTW

Whenever you write an expression that represents area or volume, check to be sure you have written the correct exponent. The exponent 2 represents two dimensions—length and width—and is used when writing a measure of area. The exponent 3 represents three dimensions—length, width, and height—and is used when writing a measure of volume.

Volume of Cylinders

We already know how to find the surface area of a cylinder by finding the sum of the area of its bases and the area of its face. To find the volume of a cylinder involves finding the area of one of its bases and multiplying that value by the height of the cylinder.

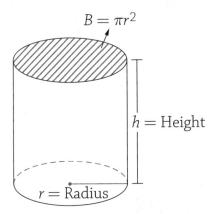

$$B = \pi r^2$$

$$h = \text{Height}$$

$$r = \text{Radius}$$

We use the following formula to find the volume of a cylinder:

$$V = \pi r^2 h$$

EXAMPLE

▶ The column of a building has a height of 15 feet and a base with a diameter of 2 feet.

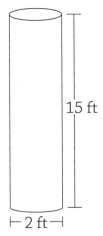

15 ft

⊢ 2 ft ⊣

▶ What is the volume of the column to the nearest tenth of a cubic foot?

▶ Use the formula for finding the volume of a cylinder: $V = \pi r^2 h$.

$$V = 3.14 \times 1^2 \times 15$$
$$= 3.14 \times 1 \times 15$$
$$= 47.1 \text{ ft}^3$$

▶ The volume of the column to the nearest tenth of a cubic foot is 47.1 ft^3.

Volume of Cones

Recall that a cone is a solid object with a circular base that tapers smoothly upward to its vertex.

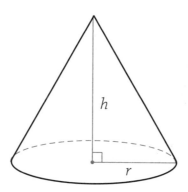

Like finding the volume of a cylinder, the formula for finding the volume of a cone involves multiplying the area of its base and its height; in other words, finding the value of $\pi r^2 h$ again. However, we must also account for the way the cone tapers upward. We use the following formula to find the volume of a cone:

$$V = \frac{1}{3}\pi r^2 h$$

▶ A sculpture in the shape of a cone has a radius of 2 feet at its base and a height of 8 feet. What is the volume of the cone to the nearest foot?

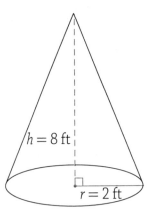

$h = 8$ ft

$r = 2$ ft

▶ Use the formula for finding the surface area of the base of the cone: $V = \frac{1}{3}\pi r^2 h.$

$$V = \frac{1}{3}\pi r^2 h$$

$$= \left(\frac{1}{3}\right)(3.14)(2^2)(8)$$

$$= \left(\frac{1}{3}\right)(3.14)(4)(8)$$

$$= \left(\frac{1}{3}\right)(100.48)$$

$$\approx 33.49$$

$$\approx 34 \text{ ft}^3$$

▶ The volume of the sculpture to the nearest foot is 34 ft³.

Volume of Pyramids

Finding the volume of a pyramid is actually a lot simpler than finding its surface area. The first thing we must do is to find the area of the pyramid's base (*B*). If we're dealing with a square pyramid, the area of the base is the same as the area of a square.

Because we are dealing with a three-dimensional solid, we know that our calculation will involve multiplying by the pyramid's height. Remember that the height (*h*) of a pyramid is not the same as its slant height (*s*). Recall that the **slant height** of a pyramid is the distance from the base of one of its triangular faces to the vertex. The **height** of a pyramid is its actual height found by measuring from a point in the center of the base to the vertex:

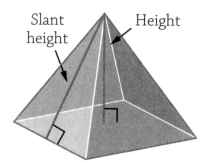

Slant height

Height

Since the pyramid tapers as it moves from the base to the vertex, the formula for finding the volume of a pyramid must accommodate this fact. To do this we multiply the product of the area of the base and height by $\frac{1}{3}$. Thus, we use the following formula to find the volume of a square pyramid:

$$V = \frac{1}{3} Bh$$

EXAMPLE

Each side of the base of a square pyramid is 8 centimeters. The pyramid has a height of 6 centimeters. What is the volume of the pyramid to the nearest centimeter?

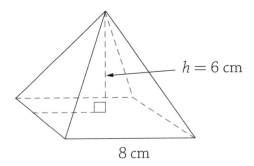

$h = 6$ cm

8 cm

▶ Use the formula for finding the area of the base of the square pyramid: $B = s^2$.

$$B = (8 \text{ cm})^2$$
$$= 64 \text{ cm}^2$$

▶ Next, plug the area of the base (B) into the formula for the volume of a square pyramid: $V = \dfrac{1}{3}(Bh)$.

$$V = \frac{1}{3}(64 \text{ cm}^2 \times 6 \text{ cm})$$

$$= \frac{1}{3}(384 \text{ cm}^3)$$

$$= 128 \text{ cm}^3$$

▶ The volume of the pyramid is 128 cm^3.

Volume of Spheres

Recall that a sphere is a solid figure that has all points on its surface that are the same distance from its center.

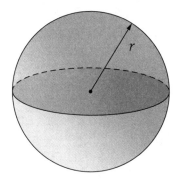

The volume of a sphere is the product of four-thirds times pi (π) times the cube of its radius (r). We use the following formula to find the volume of a sphere:

$$V = \frac{4}{3}\pi r^3$$

▶ A sphere has a diameter of 10 cm. What is the approximate volume of the sphere to the nearest centimeter?

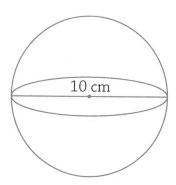

10 cm

▶ Use the formula for finding the volume of a sphere: $V = \frac{4}{3}\pi r^3$. Since the diameter of the sphere is 10 cm, its radius is 5 cm.

$$V = \frac{4}{3}\pi r^3$$

$$= \frac{4}{3}(3.14)(5 \text{ cm} \times 5 \text{ cm} \times 5 \text{ cm})$$

$$= \frac{4}{3}(3.14)(125 \text{ cm}^3)$$

$$= \frac{4}{3}(392.5 \text{ cm}^3)$$

$$= 523.\overline{33} \text{ cm}^3$$

$$\approx 523 \text{ cm}^3$$

▶ The volume of the sphere is approximately 523 cm³.

EXERCISES

EXERCISE 7-1

Find the surface area of each rectangular prism.

1. A storage box has a length of 16 inches, a width of 12 inches, and a height of 10 inches. What is the surface area of the box to the nearest inch?

2. A platform for speakers is 8 feet long, 5 feet wide, and 2 feet high. What is the surface area of the platform to the nearest foot?

EXERCISE 7-2

Find the surface area of each cylinder.

1. To the nearest tenth of a centimeter, what is the surface area of the can of tuna shown in the figure below?

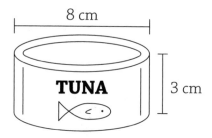

2. A scented candle in the shape of a cylinder is 5 inches tall with a radius of 2 inches. What is the surface area of the candle to the nearest inch?

EXERCISE 7-3

Find the surface area of each cone.

1. A paperweight in the shape of a cone has a diameter of 8 centimeters and a slant height of 7 centimeters. What is the surface area of the paperweight to the nearest centimeter?

2. A marble sculpture in the shape of a cone has a radius of 12 inches and a slant height of 21 inches. What is the surface area of the sculpture to the nearest inch?

EXERCISE 7-4

Find the surface area of each pyramid.

1. A paperclip dispenser has the shape of a square pyramid. Each side of the base measures 5 centimeters, and the slant height of each face is 7 centimeters. What is the surface area of the paperclip pyramid to the nearest centimeter?

2. Use the diagram that follows to find the surface area of the pyramid to the nearest foot.

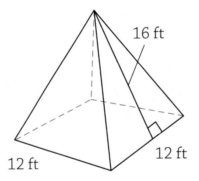

EXERCISE 7-5

Find the surface area of each sphere.

1. A bowling ball has a diameter of 22 centimeters. What is the surface area of the bowling ball to the nearest centimeter?

2. Use the diagram that follows to find the surface area of the sphere to the nearest centimeter.

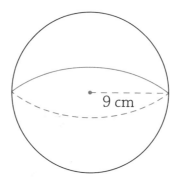

9 cm

EXERCISE 7–6

Find the volume of each rectangular prism.

1. Rolf is buying a pickup truck with a storage bed that is 6 feet long, 5 feet wide, and 2 feet high. What is the volume of the pickup truck's storage bed to the nearest foot?

2. A pedestal is in the shape of a rectangular prism. The base of the prism forms a square with sides that are 10 inches. The pedestal has a height of 30 inches. What is the volume of the pedestal to the nearest inch?

EXERCISE 7–7

Find the volume of each cylinder.

1. A container of cottage cheese is a cylinder with a diameter of 4 inches and a height of 3 inches. What is the volume of the container of cottage cheese to the nearest tenth of an inch?

2. An above-ground water storage tank is a cylinder with a diameter of 30 meters and a height of 4 meters. What is the volume of the water storage tank to the nearest meter?

EXERCISE 7–8

Find the volume of each cone.

1. Use the diagram that follows to find the volume of the cone to the nearest inch.

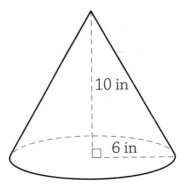

10 in

6 in

2. Based on the dimensions shown in the diagram that follows, what is the volume of the ice-cream cone to the nearest millimeter?

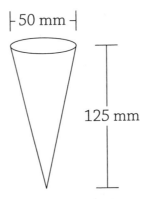

50 mm

125 mm

EXERCISE 7–9

Find the volume of each pyramid.

1. A square pyramid has a base in which each side is 10 inches. The height of the pyramid is 14 inches. What is the volume of the square pyramid to the nearest inch?

2. Based on the dimensions shown in the diagram that follows, what is the volume of the square pyramid to the nearest centimeter?

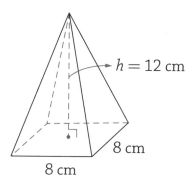

$h = 12$ cm

8 cm

8 cm

EXERCISE 7–10

Find the volume of each sphere.

1. A standard soccer ball has a radius of 11 centimeters. What is the volume of the soccer ball to the nearest centimeter?

2. Based on the diagram that follows, what is the volume of the sphere to the nearest millimeter?

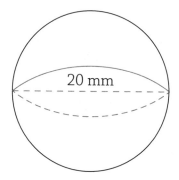

20 mm

Flashcard
App

 **Coordinate Geometry**

MUST ⚡ KNOW

⚡ Coordinate geometry is the study of the placement of points, lines, equations, and functions on the coordinate plane.

⚡ Ordered pairs of numbers consist of an *x*-coordinate (horizontal) and a *y*-coordinate (vertical).

⚡ On a coordinate plane, translations involve sliding a figure the same distance and in the same direction. Rotations involve turning a figure around a fixed point. Reflections involve flipping a figure across a specific line.

⚡ Symmetry occurs when a shape looks the same after undergoing a transformation on the coordinate plane.

⚡ A tessellation is a repeating pattern that completely covers the coordinate plane without gaps or overlaps.

I f you've ever looked at a road map or a globe, you already have some personal experience with the idea behind a coordinate system. Maps are grids that help us identify precise locations. Often the grid on a map involves some combination of letters and numbers. On a globe, the coordinate system consists of latitude and longitude coordinates that pinpoint cities, countries, and geographic features. Coordinate systems have many other practical applications—in navigation, military operations, architecture, and in satellite and Wi-Fi technology—and, yes, in geometry and algebra!

The Coordinate Plane

The **coordinate plane** is a system formed by the perpendicular intersection of a horizontal number line and a vertical number line. The horizontal number line is called the **x-axis**, and the vertical number line is called the **y-axis**. The point at which the horizontal and vertical **axes** (the plural of *axis*) meet is the **origin**, which is labeled as 0. The following grid shows the set-up of a coordinate grid.

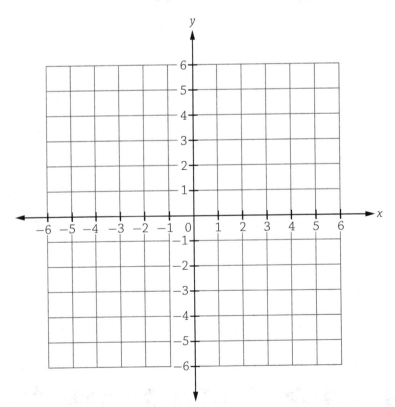

From our earlier work, in Chapter 1, we already know that on the horizontal x-axis, numbers to the right of 0 are positive and numbers to the left of 0 are negative. On the y-axis, numbers above 0 are positive and numbers below 0 are negative. Notice both axes have arrows at either end to indicate they are lines that go on forever. We also can see that 0 is the origin of both axes.

Graphing Points on the Coordinate Plane

Points on the coordinate plane are identified by **ordered pairs** of numbers that are written in parentheses, for example, $A(-2, 3)$, $B(2, 3)$, $C(2, -3)$, $D(-2, -3)$. The capital letter gives the name of the point. The first number in each ordered pair represents the position of the point on the x-axis, and the second number its position on the y-axis. By using just one pair of numbers, we can locate any point on the coordinate plane. Points A, B, C, and D are shown on the grid that follows.

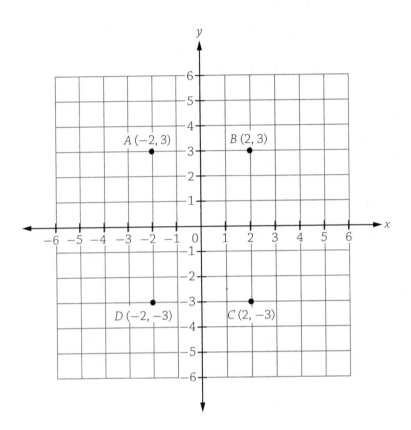

The two intersecting axes of a coordinate plane form four regions called **quadrants**. We label the quadrants I, II, III, and IV, starting with I at the upper right and moving counterclockwise. All the points located in a given quadrant have the same pattern of signs:

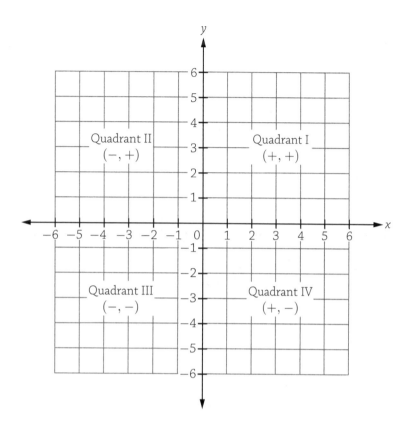

▶ Write the coordinates of points *E*, *F*, and *G* as ordered pairs and identify the quadrant each point lies in.

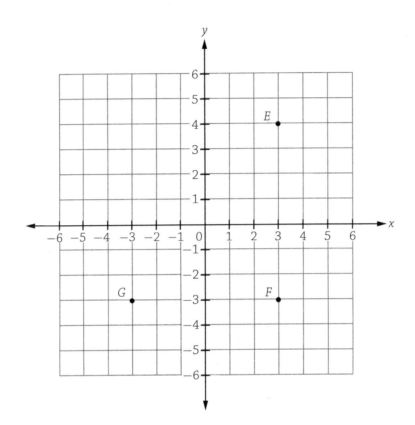

▶ To identify each point on the coordinate grid, find its x-coordinate along the horizontal axis. Find its y-coordinate along the vertical axis.

▶ Write the name of the point. Then, in parentheses, write the x-coordinate first and then the y-coordinate. Be sure to separate the coordinates with a comma.

$E(3, 4)$
$F(3, -3)$
$G(-3, -3)$

▶ Now, identify the quadrant the point lies in. Point E lies in quadrant I, point F lies in quadrant IV, and point G lies in quadrant III.

The following example asks us to identify the point that corresponds to an ordered pair of coordinates.

▶ Which points on the graph have the coordinates (–2, 4), (4, 2), (–4, –2), and (2, –4)? In which quadrant does each point lie?

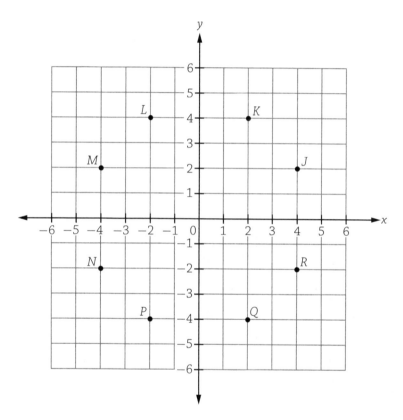

▶ For each point, use the sign of the first number in the ordered pair to locate it along the x-axis. Then, look up and down to see which point matches the y-coordinate.

▶ Point L is at (–2, 4) in quadrant II, point J is at (4, 2) in quadrant I, point N is at (–4, –2) in quadrant III, and point Q is at (2, –4) in quadrant IV.

Measuring Distance on the Coordinate Plane

We can use a coordinate graph to find the length of a line segment. The following formula gives us the distance between any two points as long as we know their coordinates.

$$d = \sqrt{(x_2 - x_1) + (y_2 - y_1)}$$

The following examples show us how to do this.

▶ See rectangle *ABCD* on the coordinate graph below. Find the length and width of the rectangle. Each unit of the graph measures 1 centimeter.

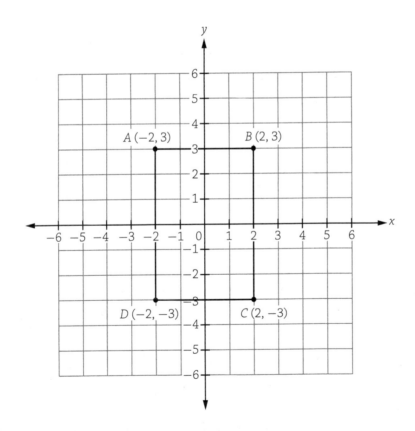

▶ First, find the width of the rectangle by using the coordinates for points A and B or points C and D. Note which point is being used: point $A = (-2, 3)$ and point $B = (2, 3)$.

$$d(\overline{AB}) = \sqrt{[2-(-2)]^2 + (3-3)^2}$$
$$= \sqrt{(4)^2 + (0)^2}$$
$$= \sqrt{16}$$
$$= 4 \text{ cm}$$

BTW
When finding the length of a line segment, it's important to keep track of which point is being used for the coordinates (x_1, y_1) and which point is being used for coordinates (x_2, y_2).

▶ Next, find the length of the rectangle using the coordinates for points A and D or points B and C. Note which point is being used: point $A = (-2, 3)$ and point $D = (-2, -3)$.

$$d(\overline{AD}) = \sqrt{[-2-(-2)]^2 + (-3-3)^2}$$
$$= \sqrt{(0)^2 + (-6)^2}$$
$$= \sqrt{36}$$
$$= 6 \text{ cm}$$

▶ Rectangle $ABCD$ has a width of 4 cm and a length of 6 cm.

Now that the width and length of rectangle $ABCD$ are known, we can use the formulas we studied in Chapter 7 to find its perimeter and area.

EXAMPLE

▶ What is the perimeter of rectangle $ABCD$? What is its area?

▶ First, apply the perimeter formula: $P = 2l + 2w$. Recall that each unit of the graph equals 1 cm.

$$P = 2l + 2w$$
$$= (2 \times 6) + (2 \times 4)$$
$$= 12 + 8$$
$$= 20 \text{ cm}$$

▶ Next, apply the area formula: $A = l \times w$.

$$A = l \times w$$
$$= 6 \times 4$$
$$= 24 \text{ cm}^2$$

▶ The rectangle has a perimeter of 20 cm and an area of 24 cm^2.

We can also use the distance formula to find the length of line segments that are not parallel to the x-axis or y-axis.

▶ What is the length of line segment $\overline{VW}$ on the following coordinate grid, to the nearest tenth? Each unit on the grid is 1 centimeter. Use the formula $d = \sqrt{(x_2 - x_1)^2 + (y_2 - y_1)^2}$.

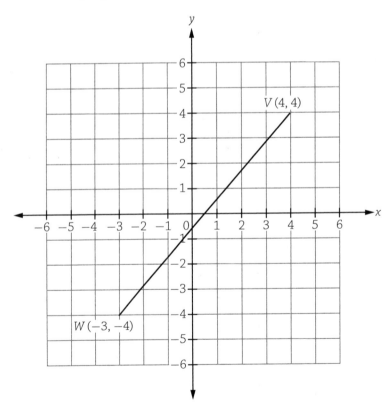

▶ Use the coordinates for point V as x_1 and y_1 and the coordinates for point W as x_2 and y_2. Plug the coordinates for each point into the formula.

$$d = \sqrt{(x_2 - x_1)^2 + (y_2 - y_1)^2}$$
$$= \sqrt{(-3 - 4)^2 + (-4 - 4)^2}$$
$$= \sqrt{(-7)^2 + (-8)^2}$$
$$= \sqrt{49 + 64}$$
$$= \sqrt{113}$$
$$\approx 10.6 \text{ cm}$$

▶ The length of line segment $\overline{VW}$ is approximately 10.6 cm.

 IRL Scientists often use a three-dimensional coordinate system that has an x-axis, a y-axis, and a z-axis. The three coordinates enable researchers to describe any location in space.

A **transformation** involves the movement of all points of a figure on the coordinate plane. The figure in its new position is referred to as the **image**. Vertices of the image are labeled the same as the original but are followed with a tick mark. Thus, A becomes A', B becomes B', and so on.

Translations

A **translation** is a transformation in which all points of the original figure are moved the same distance in the same direction. In other words, when a figure undergoes a translation, it *slides* into another position. Thus, the original figure and its image are *congruent*. In the example that follows, the original figure of $\triangle RST$ appears in quadrant II and is translated into quadrant IV.

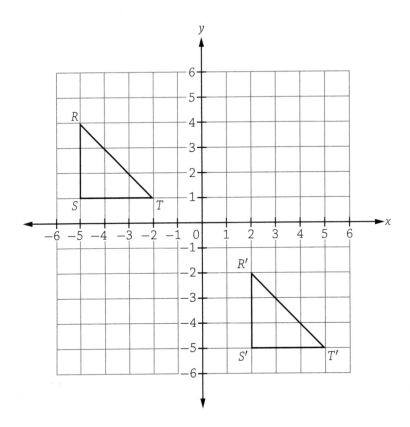

How can we describe exactly what has occurred in this translation? First, we notice that the original figure has shifted to the right. If we count the number of squares from point R to point R', we'll see that the entire triangle is shifted right 7 units. Remember that as we move right on a number line, the numbers become greater. Since every point of $\triangle RST$ has been moved along the horizontal axis the same distance in the same direction, it's reasonable to describe this shift in position as $x + 7$.

As we move up the y-axis, the numbers become greater, and as we move down the y-axis, the numbers become smaller. In this example, all the points in the original figure move down 6 spaces. We can describe this shift as $y - 6$. Both shifts (to the right and down) can be summed up neatly in **standard form** as: $(x, y) \rightarrow (x + 7, y - 6)$.

Assuming that a and b are positive, the basic rules for translating a figure into an image are:

- Slide right a units: $x \rightarrow x + a$

- Slide left a units: $x \rightarrow x - a$

- Slide up b units: $y \rightarrow y + b$

- Slide down b units: $y \rightarrow y - b$

EXAMPLE

▶ Write the coordinates of the vertices of the original figure and its image from the graph below. Describe in words the transformation that is shown. Identify the quadrants involved in the transformation. Write the standard form for this transformation.

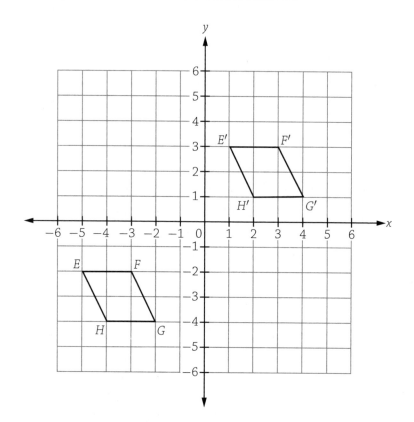

▶ First, make a table that lists the vertices of the original figure and their coordinates in the first column. In the second column, record the corresponding vertices of the image.

Original Vertices	Image Vertices
E(−5, −2)	E'(1, 3)
F(−3, −2)	F'(3, 3)
G(−2, −4)	G'(4, 1)
H(−4, −4)	H'(2, 1)

▶ The transformation, therefore, is a translation in which every point of the original figure is shifted 6 units to the right and 5 units up. The original figure is translated from quadrant III to its image in quadrant I.

▶ The standard form for this translation is $(x, y) \rightarrow (x + 6, y + 5)$.

Rotations

A **rotation** is a transformation in which all points of the original figure are turned through a given **angle of rotation** around a fixed point called the **center of the rotation**. As with translations, the original figure and its rotated image are congruent. The angle of rotation can be any number of degrees clockwise or counterclockwise. In the examples in this book, the center of rotation is *always* the origin.

In the graph below, the original figure of a pentagon is rotated 90° clockwise around the origin (0, 0) of the coordinate plane.

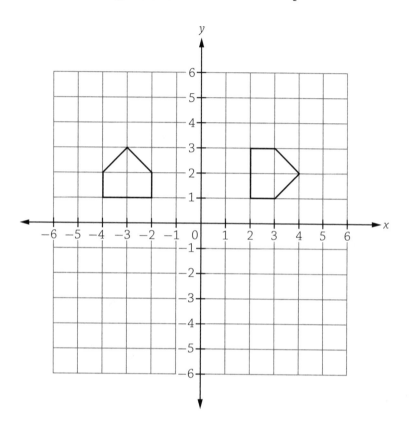

EXAMPLE

In the graph below, how many degrees is the figure rotated and in which direction?

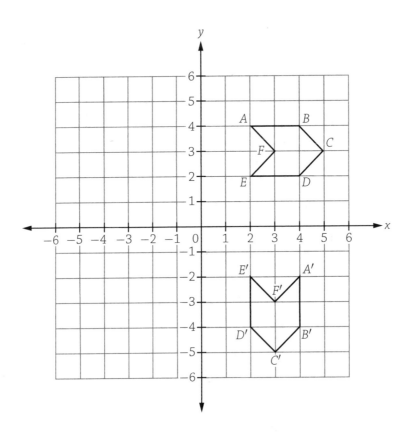

If you look at point C of the original figure, you can see that, in the image, C' is turned through a $90°$ angle, making it perpendicular to the original figure. The transformation is, therefore, a $90°$ rotation moving clockwise.

Now, let's consider the same original figure but with a different rotation.

▶ How many degrees and in which direction is the figure below rotated?

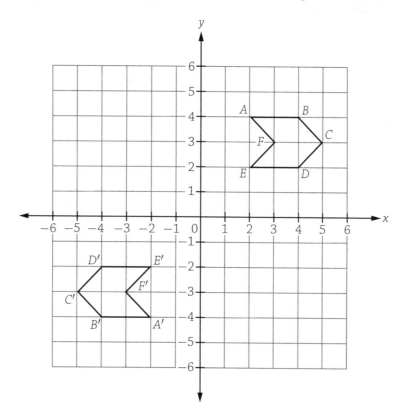

▶ Since you cannot slide the original figure in quadrant I horizontally or vertically into the position of the image, in quadrant III the image is a rotation.

▶ The position of point *A* is oriented in the opposite direction from the original figure. Points *B* and *D* have reversed their top/bottom positions in the image. Therefore, the transformation involves a rotation of 180°.

▶ Notice that this rotation could occur in a clockwise or counterclockwise direction.

These general rules will help guide us when rotating a figure:

Rotation	What to Do	Standard Form
90° Clockwise	Switch coordinates. Multiply new y-coordinate by –1.	$P(x, y) \rightarrow P'(y, -x)$
90° Counterclockwise	Switch coordinates. Multiply new x-coordinate by –1.	$P(x, y) \rightarrow P'(-y, x)$
180° Clockwise and Counterclockwise	Do not switch coordinates. Multiply both coordinates by –1.	$P(x, y) \rightarrow P'(-x, -y)$

Reflections

A **reflection** is a transformation that creates a mirror image of the original figure by flipping it over a **line of reflection**. As with translations and rotations, the original figure and its reflected image are congruent. In the example that follows, the original figure of △KLM is flipped over the x-axis to produce the reflected image of △K'L'M'.

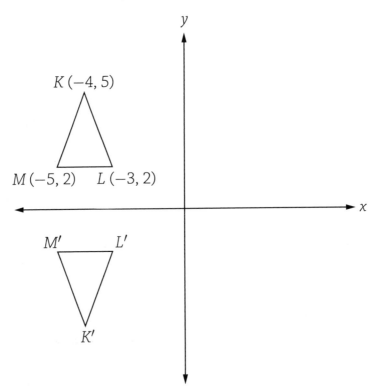

If we know the coordinates of a point on a figure, we can determine the coordinates of the same point in its reflected image. In the preceding example, we know that the vertex designated by point K in $\triangle KLM$ has coordinates of $(-4, 5)$. We don't need a grid to figure out the coordinates of the reflected point K'. To determine the coordinates of a point that has been reflected over the x-axis, multiply the y-coordinate by -1. So, point K' has the coordinates $(-4, -5)$.

We follow these rules for reflecting a figure into an image:

Reflection	What to Do	Standard Form
Over the x-axis	Multiply the y-coordinate by -1.	$P(x, y) \rightarrow P'(x, -y)$
Over the y-axis	Multiply new x-coordinate by -1.	$P(x, y) \rightarrow P'(-x, y)$

EXAMPLE

▶ When trapezoid $EFGH$, below, is reflected over the x-axis, what are the coordinates of points E', F', G', and H' in the new image?

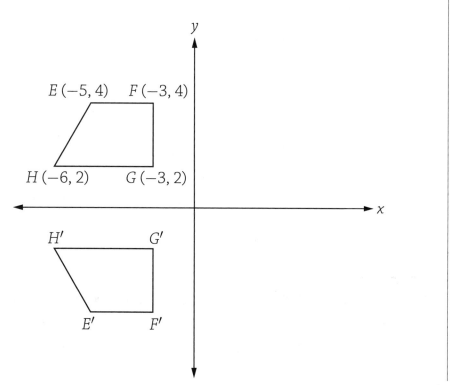

▶ Make a two-column table. Label the first column Coordinates of Original and the second Coordinates of Reflection. In the first column, record the coordinates of the vertices of the original figure.

Coordinates of Original	Coordinates of Reflection
E(–5, 4) →	
F(–3, 4) →	
G(–3, 2) →	
H(–6, 2) →	

▶ Apply the rule for determining the coordinates of a point reflected over the x-axis: multiply the y-coordinate by −1. Complete the table by recording the coordinates of the reflected points in column 2.

Coordinates of Original	Coordinates of Reflection
E(–5, 4) →	E′(–5, –4)
F(–3, 4) →	F′(–3, –4)
G(–3, 2) →	G′(–3, –2)
H(–6, 2) →	H′(–6, –2)

▶ The coordinates of the vertices of the reflected image are: $E'(-5, -4)$, $F'(-3, -4)$, $G'(-3, -2)$, and $H'(-6, -2)$.

The following example shows us how to determine the coordinates of a point when a figure is reflected over the y-axis. To find the coordinates of the reflected image, multiply the x-coordinate by −1.

EXAMPLE

▶ When trapezoid *EFGH* on the next page, is reflected over the y-axis, what are the coordinates of points E', F', G', and H' in the new image?

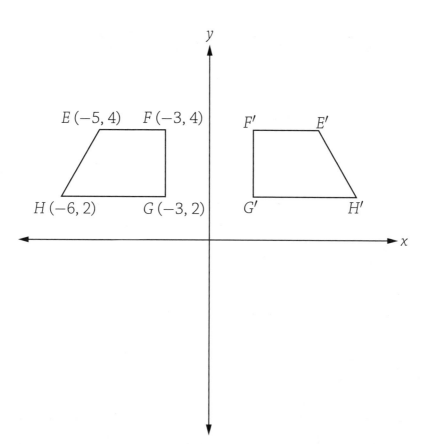

▶ Make a two-column table. Label the first column Coordinates of Original and the second-column Coordinates of Reflection. In the first column, record the coordinates of the vertices of the original figure.

Coordinates of Original	Coordinates of Reflection
E(–5, 4) →	
F(–3, 4) →	
G(–3, 2) →	
H(–6, 2) →	

▶ Apply the rule for determining the coordinates of a point reflected over the *y*-axis: multiply the *x*-coordinate by −1. Complete the table by recording the coordinates of the reflected points in column 2.

Coordinates of Original	Coordinates of Reflection
$E(-5, 4) \rightarrow$	$E'(5, 4)$
$F(-3, 4) \rightarrow$	$F'(3, 4)$
$G(-3, 2) \rightarrow$	$G'(3, 2)$
$H(-6, 2) \rightarrow$	$H'(6, 2)$

▶ The coordinates are $E'(5, 4)$, $F'(3, 4)$, $G'(3, 2)$, and $H'(6, 2)$.

Symmetry

A plane figure has **symmetry** if we can divide it into two or more congruent pieces that can be systematically arranged. In other words, we can move individual pieces of the figure without changing its shape and, therefore, the pieces of the figure remain congruent. Some of this may sound familiar. That's because it is! Transformations such as reflections, rotations, and translations are the fundamental building blocks of any study of symmetry in geometry.

Symmetry that involves reflections is often called **line symmetry**, or mirror symmetry. A plane figure has a **line of symmetry** if one or more lines can be drawn that divide the figure into two identical parts, creating mirror images of each other. Another way to think of line symmetry is to use the folding test. Ask yourself, "Can I fold this figure in such a way that one half of it completely covers the other half?" If the answer is yes, then the figure has line symmetry, and the edge created is a line of symmetry.

However, a follow-up question is in order: "Are there several ways that I can do this folding?" In fact, some regular plane figures have multiple lines of symmetry. Here are some examples of line of symmetry, from zero to five:

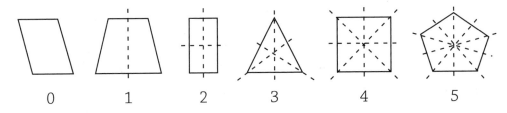

0   1   2   3   4   5

Recall that a regular polygon has sides of equal length and angles of equal measure. Of the preceding figures, the triangle, the square, and the pentagon are regular polygons. If you count the number of lines of symmetry for these three figures, you'll quickly determine that the triangle has three lines of symmetry, the square has four lines of symmetry, and the pentagon has five. In fact, the number of lines of symmetry in a regular polygon is equal to the number of sides.

EXAMPLE

▶ Copy each figure. Identify the type of polygon. Draw any line(s) of symmetry. If the polygon has no line of symmetry, write "none."

▶ The first polygon has six lines of symmetry. The one to its right has one line of symmetry. The third has no line of symmetry, and the last has two lines of symmetry.

6 lines of symmetry   1 line of symmetry

no line of symmetry   2 lines of symmetry

A plane figure has **rotational symmetry** when a turn of less than 360° produces an image that is congruent with the original. Look at the standard logo for recycling:

Notice that three identical parts form the complete design. If we add a center point and label the parts 1, 2, and 3, we will see that this design exhibits rotational symmetry. Turns of 120° clockwise or counterclockwise produce an image that exactly matches the original in size and shape. Notice that three 120° turns return the image to its original position.

Original   120° Clockwise   240° Clockwise   360° Clockwise

EXAMPLE

▶ Does the figure below exhibit symmetry? If so, what type does the figure have? Explain how you know.

▶ Determine if the figure has line symmetry by drawing lines of symmetry if they exist.

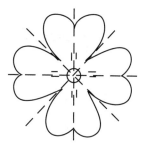

▶ Determine if the figure has rotational symmetry and, if so, its angle of rotation.

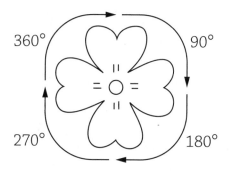

▶ The figure exhibits line symmetry with four lines of symmetry. The figure also exhibits rotational symmetry with a 90° angle of rotation.

Tessellations

A **tessellation** involves covering a plane with congruent shapes so there are no gaps or overlaps. A tessellation is often referred to as the **tiling of a plane**. In a regular tessellation, each tile is a regular polygon. In fact, there are exactly three regular polygons that tessellate a plane: squares, equilateral triangles, and regular hexagons. Each tessellation that follows is based on only one regular polygon.

Regular Tessellations

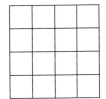

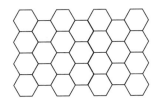

Semi-regular tessellations are composed of two or more regular polygons. The tessellation on the left is composed of hexagons, equilateral triangles, and squares. What plane figures are used in the tessellation on the right?

Semi-Regular Tessellations

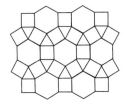
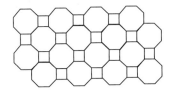

Irregular tessellations are composed of a combination of regular and irregular polygons or just one or more irregular polygons.

Irregular Tessellations

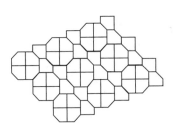
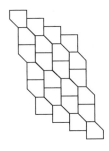

No matter which type, every true tessellation has three essential characteristics: transformations of polygons, no gaps, and no overlaps.

EXAMPLE

▶ Identify the polygon(s) used in the tessellation below. Copy the pattern and show examples of any transformation(s). What type of tessellation is shown?

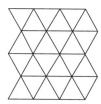

▶ This tessellation uses only equilateral triangles.

▶ Several transformations are shown in the tessellation.

Translations　　　**Reflections**　　　**Rotations**

▶ The figure is a regular tessellation that uses equilateral triangles. The tessellation involves translations, reflections, and rotations.

EXERCISES

EXERCISE 8–1

Answer each question based on the graph below.

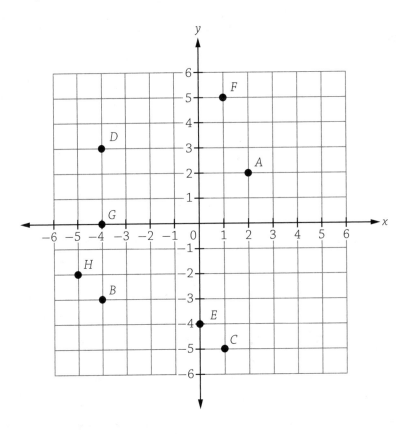

1. What are the coordinates of points *A, B, C,* and *D* on the graph? Write the ordered pairs for each point. What quadrant does each point lie in?

2. Which points on the graph have the coordinates (0, −4), (1, 5), (−4, 0), and (−5, –2)?

EXERCISE 8–2

Use the graph that follows to solve each problem. Use the distance formula in your solution. Assume each unit of the graph is equal to 1 inch.

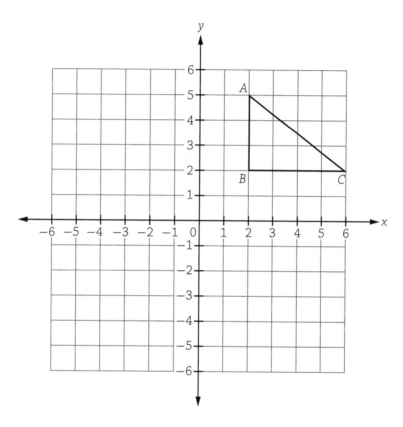

1. What is the length of line segment $\overline{BC}$?

2. What is the length of line segment $\overline{AB}$?

3. What is the length of line segment $\overline{AC}$?

4. What is the perimeter of $\triangle ABC$?

5. What is the area of $\triangle ABC$?

EXERCISE 8–3

What type of transformation is shown in the graphs of the questions below? Write translation, rotation, reflection, or tessellation.

1.

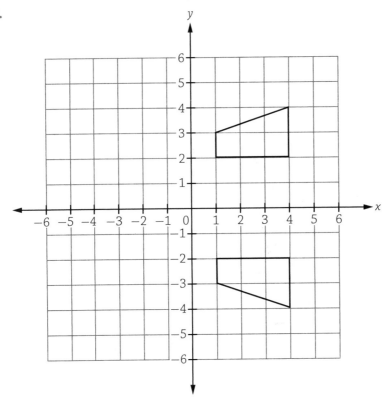

2.

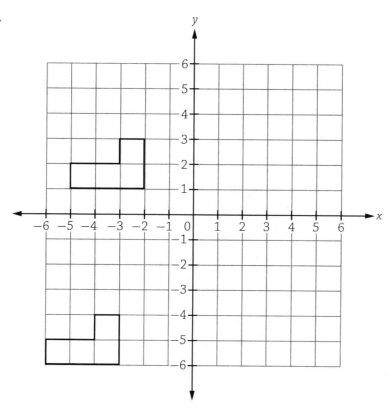

3.

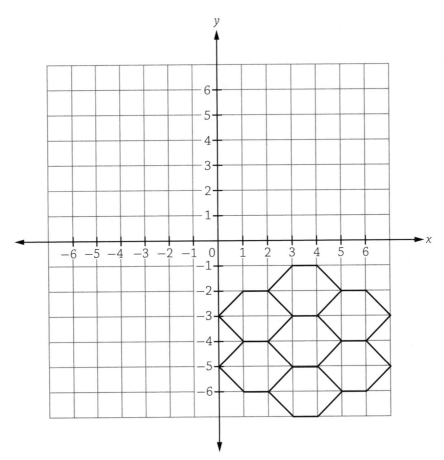

4.

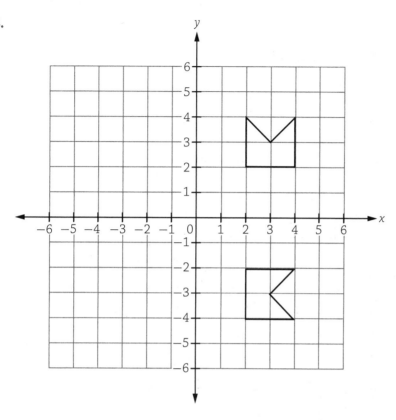

EXERCISE 8–4

Use the graph below to answer each question.

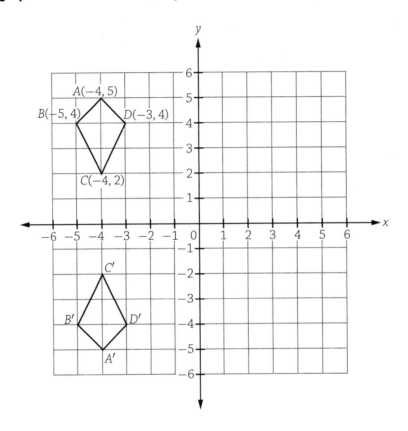

1. What type of transformation is shown?

2. What quadrant is the original figure in?

3. What quadrant is the image in?

4. What are the coordinates of the vertices A', B', C', and D'?

EXERCISE 8-5

Use the graph that follows to answer each question.

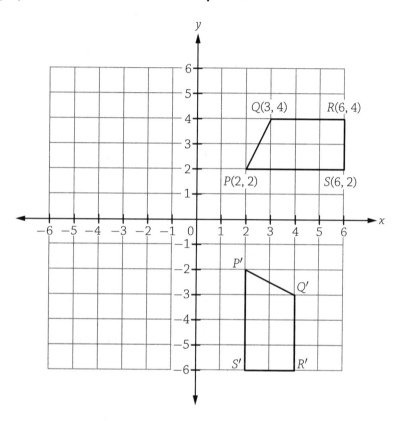

1. What type of transformation is shown?

2. What quadrant is the original figure in?

3. What quadrant is the image in?

4. What are the coordinates of the vertices *P′*, *Q′*, *R′*, and *S′*?

EXERCISE 8-6

Use the graph below to answer each of the following questions.

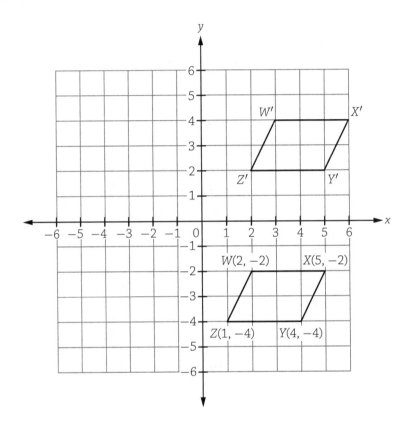

1. What type of transformation is shown?

2. What quadrant is the original figure in?

3. What quadrant is the image in?

4. What are the coordinates of the vertices W', X', Y', and Z'?

EXERCISE 8–7

Determine the number of lines of symmetry for each figure below. Copy the figure and draw the lines of symmetry, if any.

1.

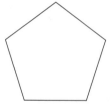

2.

3.

4.

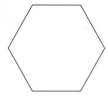

EXERCISE 8-8

For each figure, determine the type of tessellation. Where possible, identify the polygons used in the tessellation.

1.

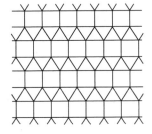

2.

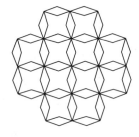

Linear Functions

MUST ⚡ KNOW

⚡ A relation is a set of ordered pairs that relates the value of an input, or *x*-coordinate, to the value of an output, or *y*-coordinate.

⚡ A function is a relation in which there is exactly one output value for each input value.

⚡ Linear equations and functions can be graphed using an input-output table, intercepts, or slope and *y*-intercept.

⚡ The slope of a line is a ratio that measures the steepness of a line by calculating its vertical change to its horizontal change.

⚡ Linear equations with the same two variables form a system of equations that can be solved by graphing, by substituting values for *x* and *y*, or by eliminating one of the variables.

⚡ Linear inequalities with two variables can be graphed on the coordinate plane.

hat are linear functions? Why are they worth studying? Linear functions have endless applications in the real world, even if you are not an engineer, a scientist, or computer programmer. For example, if you want to figure out how much money you will earn at your job over a period of time, you can set up a linear function. If you want to find out how much a car rental fee will be based on how far you travel, you can use a linear function. In fact, numerous situations in which there is an unknown quantity can be represented by linear functions. The really cool thing is you can also graph linear functions, which will enable you to find answers to your specific questions—at a glance!

Relations and Functions

Recall that an **equation** is a mathematical statement formed by equating, or placing an equal sign, between two expressions. The **solution** of an equation is the number that we substitute for a variable that makes the equation true. In other words, the unknown in an equation has a specific value. A function, however, is more general. To understand how these two concepts differ, let's begin by explaining what a relation is.

A **relation** is a set of ordered pairs of numbers that relates an input to an output. From our study of coordinate geometry in Chapter 8, we know that ordered pairs are written as two numbers placed in parentheses. For example, (3, 6) and (5, 10) are ordered pairs, with the first number representing x, or the horizontal scale of the coordinate plane, and the second number representing y, or the vertical scale. In a function, x is the input and y is the output—it's that simple!

A relation is a **function** if, and only if, for each input there is exactly only one output. In other words, the output is dependent on the input. That's why x is called the **independent variable** and y is called the **dependent variable**! In turn, the set of all input values is called the **domain** of the function. The set of all output values is called the **range** of the function.

Functions can be represented in three different ways: as a rule, a table of values, or a graph. Let's take a look at some examples.

- Rule: The output (y) is equal to the input (x) plus 2.

- Table of values:

Input (x)	−4	−2	0	2
Output (y)	−2	0	2	4

- Graph:

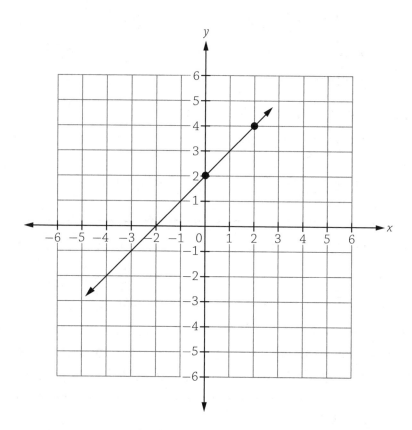

▶ Do the ordered pairs below represent a function? Explain.

$(2, 1), (3, 4), (1, -2), (0, -5)$

▶ The relation is a function. Each input has exactly one output.

Let's look at an example that presents values in a table.

▶ Is the relation represented in the table a function? Explain.

Input	-3	-3	0	3	5
Output	7	0	6	5	9

▶ The relation is not a function. The input -3 has more than one output.

The following example is a little trickier than the preceding ones.

▶ Is the relation represented by the ordered pairs, below, a function? Explain.

$(-4, 2), (-2, 2), (0, 2), (2, 2), (4, 2)$

▶ The relation is a function. Although the output for each input is always 2, each input has exactly one output. In fact, this function represents a horizontal line parallel to the x-axis at $y = 2$.

Now, let's take a look at how we can write a rule for the function represented by a list of input–output values.

EXAMPLE

▶ What function rule is represented by the following input–output values? Explain.

$$(-2, 1), (-1, 2), (0, 3), (1, 4), (2, 5)$$

▶ Compare each output value to its input value. For these ordered pairs, each output is the input value plus 3.

▶ The function rule for this set of input–output values is $y = x + 3$, or the value of y equals the value of x plus 3.

Graphing Linear Functions Using a Table

We can use the x- and y-values of an input–output table to graph a linear function.

EXAMPLE

▶ Graph the function $y = 2x - 3$. Explain what the graph shows.

▶ First, make an input–output table.

Input	-2	-1	0	1	2
Output	-7	-5	-3	-1	1

▶ Plot the ordered pairs. Draw a line through the points.

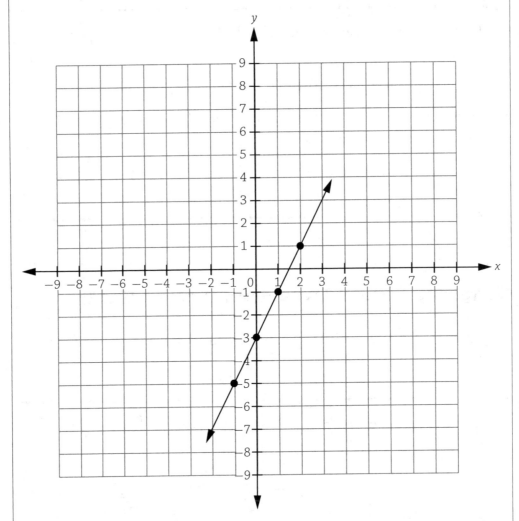

▶ The graph shows the output, or *y*-value, for each input, or *x*-value. Each point (*x*, *y*) represents a solution to the function.

Let's look at a word problem that we can solve by graphing a linear equation.

EXAMPLE

▶ The battery of Nico's computer is completely drained. If the computer takes 5 minutes to charge its battery to 10% of capacity, how long will it take for the battery to reach 70% of capacity?

▶ Make an input–output table for three points. Remember that for each 5 minutes of charging the battery's capacity increases by 10%.

Input (minutes)	5	10	15
Output (% of capacity)	10	20	30

▶ Plot the ordered pairs. Draw a line through the points.

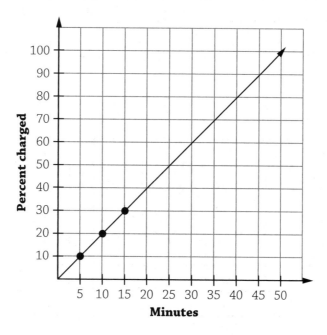

▶ The function is $y = 2x$.

▶ The x-axis tells the number of minutes of charging. The y-axis gives the percent of capacity charged. You want to know how many minutes it will take to reach a charge of 70%. Plug the numbers into the function.

$$y = 2x$$
$$70 = 2x$$
$$x = 35$$

▶ It will take 35 minutes to charge the battery to 70% of its capacity.

Intercepts

The concept of an intercept in coordinate geometry is easy to understand. The **x-intercept** is the point at which the graph of the line crosses the x-axis, and the **y-intercept** is the point at which the graph of the line crosses the y-axis. Just by looking at the graph below, we can quickly tell that the x-intercept of this linear function is −3 and the y-intercept is 2.

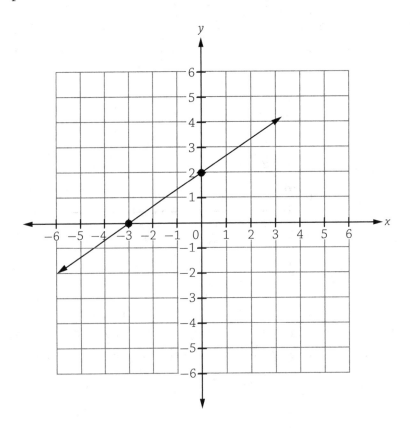

At this point, you might be wondering, "So, what's the big deal with intercepts?" Take another look at the preceding graph. What ordered pair represents the point where the line crosses the x-axis? The value of x at the x-intercept is -3, but the value of y is 0. Similarly, the value of y at the y-intercept is 2, but its x value is 0.

▶ Graph the linear function that has an x-intercept of $(2, 0)$ and a y-intercept of $(0, -5)$.

▶ Graph the two points and, simply, draw a line through them.

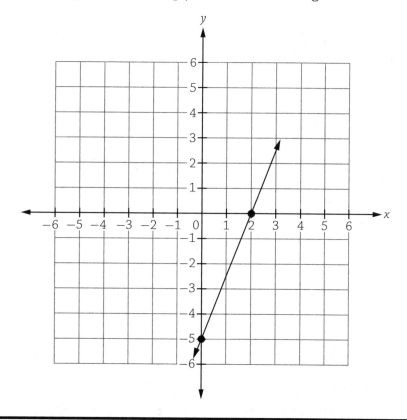

Slope-Intercept Form

A well-known method for graphing linear functions is the slope-intercept form. Slope is a measurement that represents the steepness of a line on the coordinate grid. Slope is the ratio of the vertical change of the line to its horizontal change. The terms *rise* and *run* are often used when talking about slope. The term *rise* refers to the change in y-coordinates since they are along the vertical axis. The term *run* refers to the change in the x-coordinates. In other words,

$$\text{slope} = \frac{rise}{run} = \frac{y_2 - y_1}{x_2 - x_1}$$

We only need to know the coordinates of two points on a line to determine its slope. The following example shows how to do this.

EXAMPLE

▶ A line has coordinates at points $(2, 1)$ and $(0, -3)$. What is the slope this line?

▶ Use the following formula to determine the slope.

$$\text{slope} = \frac{y_2 - y_1}{x_2 - x_1}$$
$$= \frac{-3 - 1}{0 - 2}$$
$$= \frac{-4}{-2}$$
$$= 2$$

▶ The slope of a line with points at $(2, 1)$ and $(0, -3)$ is 2.

A linear equation can be written in slope-intercept form $y = mx + b$, where m is the slope and b is the y-intercept. Knowing this not only helps us identify the slope and y-intercept of a line, it also helps us write the equation of the line, as the following example shows.

EXAMPLE

▶ A line has points at (−1, 3) and (1, −1). Determine the slope of the line. Then, graph the line.

▶ Use the formula to determine the slope.

$$\text{slope} = \frac{y_2 - y_1}{x_2 - x_1}$$

$$= \frac{-1 - 3}{1 - (-1)}$$

$$= \frac{-4}{2}$$

$$= -2$$

▶ The slope of the line is −2.

▶ Plot the two points and graph the line.

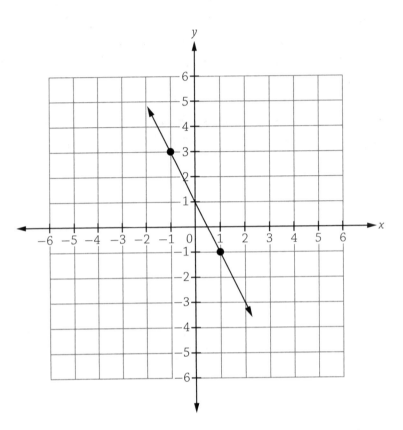

▶ Notice that this line has a negative slope and, therefore, shifts down and to the right.

Once we know the slope of a line and its *y*-intercept, it's quite easy to write an equation that represents the line. Let's see how this is done by using the slope and *y*-intercept of the preceding example.

EXAMPLE

▶ Write an equation for the line with points at $(-1, 3)$ and $(1, -1)$.

▶ From the previous Example, you know that the slope of the line is -2 and the y-intercept is 1.

▶ Substitute the values of the slope and y-intercept into the slope-intercept equation.

$$y = mx + b$$
$$y = -2x + 1$$

▶ The equation of the line with points at $(-1, 3)$ and $(1, -1)$ is $y = -2x + 1$.

Solving Systems of Equations by Graphing

Two linear equations with the same two variables form a **system of equations**. The system is described as an **independent system** if the graphs of the lines intersect at only one point. The coordinates of the point of intersection form the **solution of the system**, because the coordinates make both equations true. Let's consider an example of how to solve a system of linear equations by graphing.

EXAMPLE

▶ What is the solution of the system of equations $y = x - 1$ and $y = 2x + 2$?

▶ Graph both equations on the same coordinate plane.

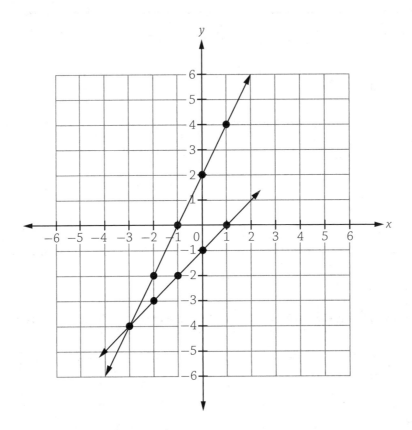

▶ The lines intersect at $(-3, -4)$.

▶ Check the solution by substituting -3 for x and -4 for y in both equations.

$$y = x - 1 \qquad\qquad y = 2x + 2$$
$$-4 = -3 - 1 \qquad\qquad -4 = 2(-3) + 2$$
$$-4 = -4 \checkmark \qquad\qquad 4 = -6 + 2$$
$$\qquad\qquad\qquad -4 = -4 \checkmark$$

▶ Since the check for both equations is a true statement $(-4 = -4)$, the solution to the system of equations $y = x - 1$ and $y = 2x + 2$ is $(-3, -4)$.

Not all pairs of equations form an independent system of equations. If the graphs of the two equations in a system never intersect, they have no solution and are called an **inconsistent system**. Let's look at an example.

▶ Graph these equations on the same coordinate grid: $y = 3x + 2$ and $y = 3x - 3$. How can we tell if the equations form an inconsistent system?

▶ Use the slope-intercept method to graph the equations since both are already in standard slope-intercept form. First, mark the y-intercept for each equation. Then, move up 3 units and 1 unit right from each intercept.

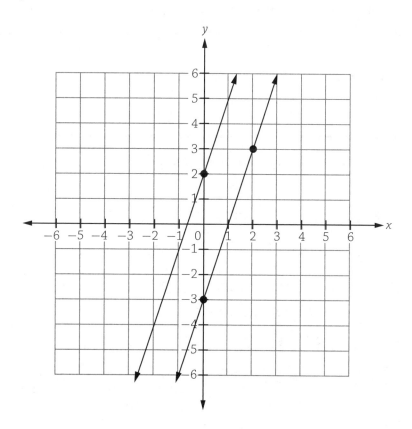

▶ We can tell by sight that the two lines in the graph are parallel and will never meet. We can also tell that they are parallel by looking at the equations and noticing that the slope of both is the same (3) and only their y-intercepts are different.

▶ Since the lines are parallel, they never interesect and have no solution. They form an inconsistent system of equations.

Sometimes when we graph two equations, they produce lines that coincide. This means that the two equations have different forms, but they are really equivalent and have exactly the same set of solutions. Two equations that are equivalent form a **dependent system**.

Here's an example of two equations that form a dependent system.

BTW

Since the two equations in a dependent system coincide at every point, the system has an infinite number of solutions.

EXAMPLE

▶ What is the solution of the system of equations $y = 2x + 3$ and $2y = 4x + 6$?

▶ First, make an input–output table for each equation.

$y = 2x + 3$

Input (x)	−1	0	1
Output (y)	1	3	5

$2y = 4x + 6$

Input (x)	−1	0	1
Output (y)	1	3	5

▶ Plot the ordered pairs. Then, draw a line through the points.

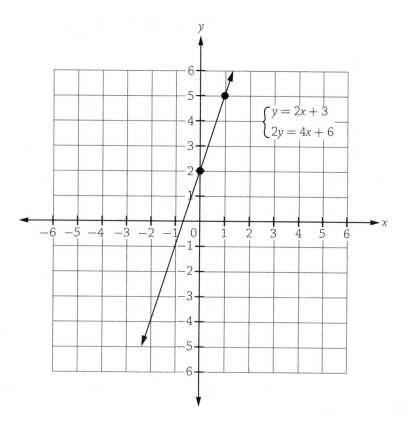

$$\begin{cases} y = 2x + 3 \\ 2y = 4x + 6 \end{cases}$$

▶ Notice that the points all fall along the same line. The two equations, therefore, form a dependent system.

The table that follows provides a summary of the key points to remember about the different kinds of systems of linear equations.

Type of System	Intersection(s)	Solution(s)
Independent	One intersection	One solution
Inconsistent	No intersection	No solutions
Dependent	Coincides	Infinite solutions

Solving Systems of Equations by Substitution

Graphing equations is not the only way to solve a system of linear equations. We can apply our knowledge of algebra and use substitution. In substitution, the first step is to isolate one of the variables (either x or y) in one of the equations. Then, we substitute the expression found in the first step into the other equation and solve. Once we have a value for the first variable, we substitute that value into the first equation and solve for the other variable.

This process may sound complex but, if completed carefully, substitution is a much quicker way than graphing to find a solution to a system of linear equations. We can decide for ourselves by examining the example that follows.

EXAMPLE

▶ What is the solution of this system of equations?

Equation 1: $\qquad x + 2y = 5$

Equation 2: $\qquad 2x - 3y = 3$

▶ Isolate a variable (either x or y) in one of the two equations. Isolate x in equation 1.

$$x + 2y = 5$$
$$x + 2y - 2y = 5 - 2y$$
$$x = -2y + 5$$

▶ Substitute the expression for the variable you just found in the previous step into equation 2 and then solve.

$$2x - 3y = 3$$
$$2(-2y + 5) - 3y = 3$$
$$-4y + 10 - 3y = 3$$
$$(-4y - 3y) + 10 = 3$$
$$-7y + 10 = 3$$
$$-7y + 10 - 10 = 3 - 10$$
$$\frac{-7y}{-7} = \frac{-7}{-7}$$
$$y = 1$$

▶ Substitute the value for the variable in the previous step into equation 1 and solve.

$$x + 2y = 5$$
$$x + 2(1) = 5$$
$$x + 2 - 2 = 5 - 2$$
$$x = 3$$

▶ x equals 3 and y equals 1.

▶ Check the solution by substituting 3 for x and 1 for y in both equations.

Equation 1	Equation 2
$x + 2y = 5$	$2x - 3y = 3$
$3 + 2(1) = 5$	$2(3) - 3(1) = 3$
$5 = 5$	$6 - 3 = 3$
	$3 = 3$

▶ Since the check for each equation produces a true statement—that is, $5 = 5$ and $3 = 3$—the solution to the system of equations $x + 2y = 5$ and $2x - 3y = 3$ is $(3, -1)$.

Solving Systems of Equations by Elimination

Another method for solving systems of equations is by eliminating one of the variables. We can do this by adding or subtracting one equation from the other in order to make one of the variables drop out.

EXAMPLE

▶ What is the solution of this system of equations?

Equation 1: $3x + y = -7$
Equation 2: $5x - y = -9$

▶ Notice that the x-terms have coefficients that are opposites. Therefore, we can eliminate the x-variable by addition.

$$3x + y = -7$$
$$+ \quad 5x - y = -9$$
$$\overline{ \quad 8x = -16}$$

▶ Solve the resulting equation for x by dividing both sides by 8.

$$8x = -16$$
$$\frac{8x}{8} = \frac{-16}{8}$$
$$x = -2$$

▶ Substitute the value of x into one of the original equations and solve for y.

$$3x + y = -7$$
$$3(-2) + y = -7$$
$$-6 + y = -7$$
$$-6 + 6 + y = -7 + 6$$
$$y = -1$$

▶ Check by substituting the values for x and y into one or both of the original equations.

$$5x - y = -9$$
$$5(-2) - (-1) = -9$$
$$-10 - (-1) = -9$$
$$-9 = -9$$

$$3x + y = -7$$
$$3(-2) + (-1) = -7$$
$$-6 + (-1) = -7$$
$$-7 = -7$$

▶ Since the check for both equations produces true statements—that is, $-9 = -9$ and $-7 = -7$—the solution to the system of equations $3x + y = -7$ and $5x - y = -9$ is $(-2, -1)$.

In some problems, we also may need to multiply or divide one of the equations by a constant in order to eliminate one of the variables. The key thing to remember is that what we do to one side of the equation we must also do to the other side.

▶ What is the solution of this system of equations?

Equation 1: $2x + y = 11$
Equation 2: $x + 3y = 18$

▶ First, multiply both sides of equation 2 by -2.

$$-2(x + 3y) = -2(18)$$
$$-2x - 6y = -36$$

▶ Now the pair of equations looks like this:

Equation 1: $2x + y = 11$
Equation 2: $-2x - 6y = -36$

▶ Next, add the equations. Notice that the y terms cancel out.

$$2x + y = 11$$
$$+ \quad -2x - 6y = -36$$
$$\overline{ -5y = -25}$$

▶ Solve for x by dividing both sides of the new equation by -5.

$$\frac{-5y}{-5} = \frac{-25}{-5}$$
$$y = 5$$

▶ Substitute the value of y into one of the original equations and solve for x.

$$2x + y = 11$$
$$2x + 5 = 11$$
$$2x + 5 - 5 = 11 - 5$$
$$2x = 6$$
$$\frac{2x}{2} = \frac{6}{2}$$
$$x = 3$$

▶ Check by substituting the values for x and y into one or both of the original equations.

$$2x + y = 11$$
$$2(3) + 5 = 11$$
$$6 + 5 = 11$$
$$11 = 11 \checkmark$$

$$x + 3y = 18$$
$$3 + 3(5) = 18$$
$$3 + 15 = 18$$
$$18 = 18 \checkmark$$

▶ Since the checks for both equations produces true statements—that is, $11 = 11$ and $18 = 18$—the solution to the system of equations $2x + y = 11$ and $x + 3y = 18$ is $(3, 5)$.

Now, let's look at a word problem that involves finding the solution of a linear system of equations.

EXAMPLE

▶ Sara is deciding whether she should buy skiing equipment or continue to rent the equipment. She pays $50 each day she skis and rents the equipment. If she purchases the equipment, it costs her $600 plus $25 each day she skis. How many times must she rent skiing equipment for it to cost the same as buying the equipment?

▶ Write an equation that shows how to find Sara's cost for renting the skiing equipment based on the number of times she skis.

Equation 1 $y =$ the total cost if Sara rents the equipment
                                      $x =$ the number of rentals

$\rightarrow y = 50x$

▶ Then, write an equation that shows how to find Sara's cost for buying the skiing equipment based on the number of times she skis.

Equation 2 $y =$ the total cost if Sara buys the equipment
                                        $x =$ the number of days she skis

$\rightarrow y = 25x + 600$

▶ Substitute the value for y in equation 1 into equation 2. Solve for x.

$$y = 25x + 600$$
$$50x = 25x + 600$$
$$50x - 25x = 25x - 25x + 600$$
$$25x = 600$$
$$\frac{25x}{25} = \frac{600}{25}$$
$$x = 24$$

▶ Substitute the value for $x = 24$ into equation 1 and solve for y.

$$y = 50x$$
$$y = 50(24)$$
$$y = 1,200$$

▶ Check the solution by substituting 24 for x and 1,200 for y in both equations.

Equation 1

$$y = 50x$$
$$1,200 = 50 \times 24$$
$$1,200 = 1,200$$

Equation 2

$$y = 25x + 600$$
$$1,200 = (25 \times 24) + 600$$
$$1,200 = 600 + 600$$
$$1,200 = 1,200$$

▶ Since the checks for both equations produce true statements—that is, $1,200 = 1,200$—Sara must go skiing and rent equipment 24 times for it to cost the same as buying the skiing equipment and skiing 24 times.

Linear Inequalities with Two Variables

In Chapter 4, we studied how to solve inequalities by graphing them on a number line. However, those inequalities involved only one variable. To solve linear inequalities with two variables, we graph them on a coordinate plane in much the same way that we graph a linear equation. The graph of a linear inequality with two variables is represented by a dashed or solid line that separates the grid into two **half planes**. The section of the plane that is shaded includes all the solutions of the inequality. The following example shows us how to do this.

EXAMPLE

▶ Graph the inequality $y - 2x > 4$. Identify the half plane that contains the solutions to the inequality.

▶ For convenience, we change the greater than sign ($>$) to an equal sign ($=$) and then write the equation in slope-intercept form ($y = mx + b$).

$$y - 2x > 4 \rightarrow y - 2x = 4$$

$$y - 2x = 4$$
$$y - 2x + 2x = 4 + 2x$$
$$y = 2x + 4$$

▶ Graph the inequality.

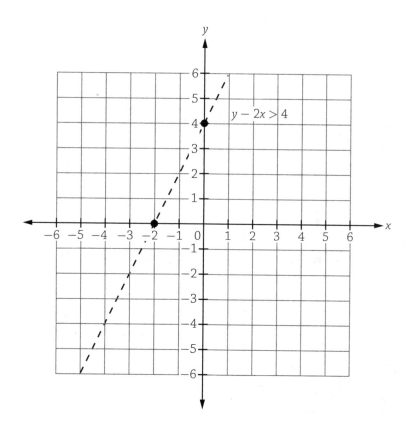

▶ Use (0, 0) as a test point to determine which part of the half plane contains the solutions.

$$y - 2x > 4$$
$$\downarrow$$
$$0 - 2(0) > 4$$
$$0 \not> 4$$

BTW

When checking the solutions of a linear inequality, it's often convenient to substitute (0, 0) for the values of the x-variables and y-variables. In fact, we can use any point to test the inequality's solutions.

▶ Therefore, (0, 0) is not a solution to the inequality $y - 2x > 4$.

▶ Shade the half plane that contains the solutions to the inequality. Since (0, 0) is *not* a solution, the other half plane, which doesn't contain this point, is shaded.

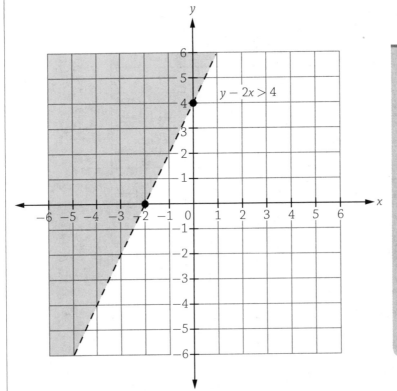

BTW

Notice that in this graph a dashed line is used to separate the two half planes. A dashed is used to show that the points on the line are not solutions to the inequality. When points on the line are solutions, a solid line is used to separate the two half planes.

Use a dashed line when the inequality contains the symbol > or <. Use a solid line line when the inequality contains the symbol ≥ or ≤.

EXERCISES

EXERCISE 9–1

Indicate whether the ordered pairs in each question below represent a function and explain your answer.

1. $(-2, 3), (2, -3), (4, 5), (4, 8)$

2. $(-4, 2), (-2, 0), (0, 2), (2, 4)$

3.

Input (x)	−2	−1	0	1	2
Output (y)	4	4	4	4	4

4.

Input (x)	−5	0	5	5	10
Output (y)	−3	2	7	−7	12

EXERCISE 9–2

Write a function rule that represents the values in the following input–output values.

1. $(-2, -4), (-1, -3), (0, -2), (1, -1), (2, 0)$

2. $(-2, 3), (-1, 4), (0, 5), (1, 6), (2, 7)$

3.

Input (x)	−2	−1	0	1	2
Output (y)	−5	−4	−3	−2	−1

4.

Input (x)	−2	−1	0	1	2
Output (y)	2	3	4	5	6

EXERCISE 9–3

Complete the input–output table for each function and then graph it.

1. $y = 2x + 2$

Input (x)	−2	−1	0	1	2
Output (y)	?	0	2	?	6

2. $y = 2x - 2$

Input (x)	−2	−1	0	1	2
Output (y)	?	−4	?	0	?

EXERCISE 9–4

Graph each linear function based on the given x-intercept and y-intercept.

1. x-intercept $= (-2, 0)$
   y-intercept $= (0, 3)$

2. x-intercept $= (4, 0)$
   y-intercept $= (0, 2)$

EXERCISE 9–5

Determine the slope of the line with the points shown below.

1. $(2, 1)$ and $(3, 4)$

2. $(2, 6)$ and $(3, 2)$

3. $(4, -3)$ and $(7, 3)$

4. $(1, 4)$ and $(3, -6)$

EXERCISE 9–6

Determine the slope of each line based on the coordinates of the points. Then, graph the line, identify the y-intercept, and write an equation for the line.

1. $(-1, 2)$ and $(-3, -2)$

2. $(-1, 4)$ and $(2, -5)$

EXERCISE 9–7

Graph the system of equations. Write the solution and whether the system is independent, inconsistent, or dependent.

1. $y = x + 2$ and $y = -x + 2$

2. $y = 2x - 1$ and $2y = 4x - 2$

3. $y = -2x + 2$ and $y = x + 5$

4. $y = 4x + 1$ and $y = 4x - 3$

EXERCISE 9–8

Solve each system of equations by substitution or elimination. Check the solution.

1. $x = y - 5$ and $2x + y = -4$

2. $x + 4y = 4$ and $x - y = -6$

3. $y + 2x = 9$ and $y = 2x - 3$

4. $y = 6x + 4$ and $2y = -6x - 10$

EXERCISE 9–9

Solve each linear inequality by graphing and then check the solution by using a test point.

1. $y - 2x > 3$

2. $y \leq -2x + 3$

Polynomials

MUST KNOW

⚡ A polynomial consists of one or more terms that contain a number and a variable or the product of a number and several variables.

⚡ A polynomial is in standard form when the exponents of its terms are written in descending order.

⚡ Like terms in polynomials can be added and subtracted.

⚡ Polynomials can be multiplied by other polynomials, using the distributive property and the rules for exponents.

⚡ Polynomials can be divided by monomials.

ave you ever taken a ride on a roller coaster and wondered how the cars stay on the tracks instead of flying off? Have you ever taken a ride in a car along a very winding road and thought about who made certain the road was safe to drive? The answer to both questions is engineers—engineers who knew all about curves. Just as linear functions represent straight lines, polynomials represent curved lines.

Engineering is not the only profession that uses polynomials. Economists use them to figure out how the economy is growing, medical researchers use them to determine the way bacteria and viruses grow and spread, and financiers use them to calculate people's net worth. Even customers shopping in supermarkets may use polynomials in their heads to estimate the total cost of their purchases.

It's important to understand what a monomial is before trying to define a polynomial. A **monomial** is an algebraic expression consisting of one term that is a number, a variable, or the product of a number and one or more variables. Examples of monomials include expressions such as 5, x, $10y$, $4x^2$, $3xy$, and $2x^2y$.

A **polynomial** is an algebraic expression that represents a monomial or the sum of two or more monomials. Polynomials are often classified by the number of terms they contain. The table below shows examples of some common types of polynomials

Name of Polynomial	Number of Terms	Examples
monomial	1	$7, 6x, 3xy, 5x^2y$
binomial	2	$4x + 2, 5y - 3, 2xy + 8$
trinomial	3	$2x^2 + 9x + 10, 8y^3 - 6y^2 - 6$

Notice that a polynomial is sometimes written as subtraction rather than addition, as we can see in the examples $5y - 3$ and $8y^3 - 6y^2 - 6$. We really should think of such subtraction expressions as the addition of a negative number or term: that is, $5y - 3$ really means $5y + -3$.

Although polynomials such as monomials, binomials, and trinomials are common, in fact, polynomials can have any number of terms.

Polynomials are usually written in standard order, meaning the term with the highest exponent comes first, the next highest comes second, and so on. The **degree** of a polynomial refers to the value of the greatest exponent in any of its terms. Thus, the polynomial $8y^3 - 6y^2 - 6$ has a degree of 3, whereas $2x^2 + 9x + 10$ has a degree of 2.

EXAMPLE

▶ Write the polynomial $6x + 4x^2 + 8x^3 + 24$ in standard order. Identify the degree of the polynomial.

▶ Identify the exponent of each term. Write the terms of the polynomial in descending order from the term with the greatest exponent to the one with the smallest exponent.

$$8x^3$$
$$4x^2$$
$$6x$$
$$24$$

▶ The standard form of this polynomial is $8x^3 + 4x^2 + 6x + 24$. Since the greatest exponent is 3, the polynomial has a degree of 3.

Sometimes a polynomial must be simplified by combining like terms before writing it in standard form. **Like terms** have variables and exponents that are the same.

EXAMPLE

▶ Simplify the polynomial that follows, write it in standard form, and identify its degree.

$$3y^2 + 8y - y^2 - 5y^3 - 4y + 12$$

▶ Combine like terms.

$$(3y^2 - y^2) + (8y - 4y) + (-5y^3) + 12$$
$$= (2y^2) + (4y) + (-5y^3) + 12$$

▶ Write the terms of the polynomial in descending order from the term with the greatest exponent to the one with the least exponent.

$$-5y^3 + 2y^2 + 4y + 12$$

▶ The simplified standard form of the polynomial is $-5y^3 + 2y^2 + 4y + 12$. Since the greatest exponent is 3, the polynomial has a degree of 3.

Adding Polynomials

We can add polynomials simply by adding like terms. To do this, we must be careful to align like terms when we set up the problem.

EXAMPLE

▶ What is the sum of $3x^3 + 5x^2 - 8 + 6x$ and $2x - x^2 + 2x^3 - 4$?

▶ Write each polynomial in standard form.

$$3x^3 + 5x^2 - 8 + 6x \qquad \rightarrow 3x^3 + 5x^2 + 6x - 8$$
$$2x - x^2 + 2x^3 - 4 \qquad \rightarrow 2x^3 - x^2 + 2x - 4$$

▶ Set up the problem in vertical addition format. Align the terms.

$$\begin{array}{r} 3x^3 + 5x^2 + 6x - 8 \\ + \ 2x^3 - \ x^2 + 2x - 4 \\ \hline \end{array}$$

▶ Add like terms.

$$3x^3 + 5x^2 + 6x - 8$$
$$+ \ 2x^3 - \ x^2 + 2x - 4$$
$$\overline{\ 5x^3 + 4x^2 + 8x - 12}$$

▶ The sum of these polynomials written in standard form is $5x^3 + 4x^2 + 8x - 12$.

Sometimes, one of the polynomials we are adding may have a term with a particular exponent that the other polynomial does not have. In this case, we can still add the polynomials, but we must be especially careful to align the terms correctly. An example of how to handle this situation follows.

EXAMPLE

▶ What is the sum of $2x^3 - x + 4x^2 - 10$ and $4x^3 + 6 + 3x^2$?

▶ Write each polynomial in standard form.

$$2x^3 - x + 4x^2 - 10 \qquad \to 2x^3 + 4x^2 - x - 10$$
$$4x^3 + 6 + 3x^2 \qquad \to 4x^3 + 3x^2 + 6$$

▶ Set up the problem in vertical addition format.

$$2x^3 + 4x^2 - x - 10$$
$$+ \ 4x^3 + 3x^2 \qquad + 6$$

▶ Add like terms.

$$2x^3 + 4x^2 - x - 10$$
$$+ \ 4x^3 + 3x^2 \qquad + 6$$
$$\overline{\ 6x^3 + 7x^2 - x - 4}$$

▶ The sum of these polynomials, written in standard form, is $6x^3 + 7x^2 - x - 4$.

Subtracting Polynomials

Recall that subtraction is the inverse (opposite) of addition. This means that we can think of subtraction as the addition of the inverse of a polynomial. Study how this is done in the following example.

EXAMPLE

▶ What is the difference between $-2x + 5x^2 + 3$ and $2x^2 - 8 + 4x$?

▶ Write each polynomial in standard form.

$$-2x + 5x^2 + 3 \qquad \rightarrow 5x^2 - 2x + 3$$
$$2x^2 - 8 \ + 4x \qquad \rightarrow 2x^2 + 4x - 8$$

▶ Set up the problem in vertical format. Remember that, to subtract a polynomial, you add its inverse.

$$5x^2 - 2x + 3$$
$$+ \ -(2x^2 + 4x - 8)$$

▶ Add like terms.

$$5x^2 - 2x + 3$$
$$+ \ -2x^2 - 4x + 8$$
$$\overline{ \ 3x^2 - 6x + 11}$$

▶ The difference between these polynomials, written in standard form, is $3x^2 - 6x + 11$.

As with the addition of polynomials, if one term in a subtraction problem is missing, just leave a space for it when setting up the problem in vertical format.

▶ What is the difference between the polynomials $3y^3 + 2y^2 - 4y + 5$ and $y^3 + 4y^2 - 3$?

▶ Since the polynomials are already in standard form, begin by setting up the problem in vertical format. Remember that, to subtract a polynomial, you add its inverse.

$$
\begin{array}{r}
3y^3 + 2y^2 - 4y + 5 \\
+ \ -(y^3 + 4y^2 \quad\ \ - 3) \\
\hline
\end{array}
$$

▶ Add like terms.

$$
\begin{array}{r}
3y^3 + 2y^2 - 4y + 5 \\
+ \ -y^3 - 4y^2 \quad\ \ + 3 \\
\hline
2y^3 - 2y^2 - 4y + 8
\end{array}
$$

▶ The difference between these polynomials, written in standard form, is $2y^3 - 2y^2 - 4y + 8$.

Multiplying Monomials and Polynomials

Recall that the commutative property of addition states that the order in which factors are multiplied does not change the product. By the associative property, we also know that the way factors are grouped does not affect the product. Both these properties come into play when multiplying polynomials. Let's look at the simplest case, that of multiplying two monomials.

▶ What is the product of $4x^2$ and $-5x$?

▶ Begin by expanding the expression into its components: $4x^2 \times -5x = 4 \times x^2 \times -5 \times x$.

▶ Regroup the factors: $4 \times -5 \times x^2 \times x$.

▶ Multiply the coefficients: $-20 \times x^2 \times x$.

▶ Apply the product of powers property: $-20x^{2+1} = -20x^3$.

▶ The product of $4x^2$ and $-5x$ equals $-20x^3$.

We can also use the distributive property and the rules of exponents to find the product of polynomials.

EXAMPLE

▶ What is the product of $3y(2y^2 - 4)$?

▶ Use the distributive property.

$$3y(2y^2 - 4) = (3y \times 2y^2) + (3y \times -4)$$

▶ Apply the product rules for exponents and the commutative property for multiplication: $6y^3 - 12y$.

Finding the product of polynomials sometimes involves simplifying the power of a product.

EXAMPLE

▶ Simplify the polynomial expression $t(-2t^2 - 4)$.

▶ Use the distributive property: $t(-2t^2 - 4) = (t \times -2t^2) + (t \times -4)$.

▶ Apply the product rules for exponents and the commutative property for multiplication.

$$= -2t^3 + -4t$$
$$= -2t^3 - 4t$$

In some problems, finding the product of two polynomials involves simplifying the power of a power.

▶ Simplify the polynomial expression $(4t^3)^2$.

▶ Apply the power of a product property: $(4t^3)^2 = 4^2 \times (t^3)^2$.

▶ Apply the power of a power property: $16 \times t^{3 \times 2}$.

▶ Simplify: $16t^6$.

Multiplying Binomials

All of the problems we have dealt with in the preceding section involve multiplying monomials and polynomials. Problems in algebra often ask another question, "How do we find the product of two binomials?"

One common method for finding the product of two binomials is to make a table. We can see how this is done in the example that follows.

▶ What is the product of $4x + 3$ and $-6x + 2$?

▶ Create a table. Write the first binomial on the left side of the table and the second binomial at the top of the table.

	$-6x$	2
$4x$		
3		

▶ Use multiplication to complete the table.

	$-6x$	2
$4x$	$-24x^2$	$8x$
3	$-18x$	6

▶ Write the product: $-24x^2 + 8x - 18x + 6$.

▶ Simplify the product by combining like terms: $-24x^2 - 10x + 6$.

Another method for multiplying binomials is to set the problem as a vertical multiplication problem.

EXAMPLE

▶ What is the product of $2x - 3$ and $2x + 4$?

▶ Write the problem in a vertical multiplication format.

$$2x - 3$$
$$\times \ 2x + 4$$
$$\overline{}$$

▶ Multiply 4 by $2x - 3$ and then multiply $2x$ by $2x - 3$. Be sure to line up like terms.

$$2x - 3$$
$$\times \qquad 2x + 4$$
$$\overline{}$$
$$8x - 12$$
$$4x^2 - 6x$$
$$\overline{}$$

▶ Add the partial products.

$$2x - 3$$
$$\times \qquad 2x + 4$$
$$\overline{}$$
$$8x - 12$$
$$4x^2 - 6x$$
$$\overline{}$$
$$4x^2 + 2x - 12$$

▶ The product of these binomials is $4x^2 + 2x - 12$.

The most efficient method for multiplying binomials is called the **FOIL** method. Each letter tells us the operation we should perform and in what order. Here's how to use the FOIL method. The FOIL method is the sum of multiplying the:

first terms first

↓

outer terms next

↓

inner terms next

↓

last terms last

In the language of algebra, this means: $(a + b)(c + d) = ac + ad + bc + bd$.

Let's see how the FOIL method can be used to find the product of two binomials.

EXAMPLE

▶ What is the product of $3w - 2$ and $5w + 7$?

▶ Multiply the *first* terms.

$$(3w - 2)(5w + 7) \rightarrow 3w \times 5w = 15w^2$$

▶ Multiply the *outer* terms.

$$(3w - 2)(5w + 7) \rightarrow 3w \times 7 = 21w$$

▶ Multiply the *inner* terms.

$$(3w - 2)(5w + 7) \rightarrow -2 \times 5w = -10w$$

▶ Multiply the *last* terms.

$$(3w - 2)(5w + 7) \rightarrow -2 \times 7 = -14$$

▶ Write all the products in order.

$$15w^2 + 21w - 10w - 14$$

▶ Combine like terms and write the product in standard form.

$$15w^2 + 11w - 14$$

Dividing Polynomials by Monomials

Polynomials can also be divided by monomials. Here's an example of how we can do this.

▶ What is the quotient of $6x^2 - 8x + 4$ divided by $2x$?

▶ Write the problem as a rational expression.

$$\frac{6x^2 - 8x + 4}{2x}$$

▶ Break up the problem into its components by placing each term in the numerator over the denominator.

$$\frac{6x^2}{2x} - \frac{8x}{2x} + \frac{4}{2x}$$

▶ Simplify each term, using the rules for exponents and reducing fractions to simplest form.

$$3x - 4 + \frac{4}{2x}$$

$$3x - 4 + \frac{2}{x}$$

▶ The quotient of $6x^2 - 8x + 4$ and $2x$ is $3x - 4 + \frac{2}{x}$.

EXERCISES

EXERCISE 10–1

Simplify and write each polynomial in standard form. Identify the degree of the polynomial.

1. $6x^3 - 4x + 8x^3 - 6$

2. $12 + 7t^2 + 4 - 2t^2 + 4t$

3. $-2 + 4n^2 - 8n + 6n^2$

4. $9c + 7c^4 + 10 - 6c^4 + 2c^2 - 2$

EXERCISE 10–2

Find the sum of the polynomials.

1. $4y^2 - 2y + 6$ and $2y^2 + 5y - 2$

2. $5z^2 - 4z + 3$ and $3z^2 + 1$

3. $2s^2 + 7s + 5$ and $4s^2 - 6s - 3$

4. $3x^2 - 3x + 4$ and $9x^2 - 8$

EXERCISE 10–3

Find the difference between the polynomials.

1. $8w^2 - 4$ and $2w^2 + 3$

2. $5a^2 - 6a + 8$ and $4a^2 - a + 6$

3. $9t^2 + 5t - 7$ and $5t^2 - 4t$

4. $-4x^2 + 8x + 5$ and $2x^2 - 4x - 3$

EXERCISE 10–4

Find the product of the polynomials.

1. $5x$ and $4x^2 - 2$

2. $3y$ and $-2y^2 + 3$

3. $(5s^3)^2$

4. $4t$ and $5t^2 + 4$

EXERCISE 10–5

Find the product of the binomials using the FOIL method.

1. $(x + 6)(2x - 2)$

2. $(n^2 - 4)(n^2 + 2)$

3. $(2w - 5)(3w + 7)$

4. $(5a^2 - 3)(-2a + 3)$

EXERCISE 10–6

Find the quotient of the polynomials.

1. $(8y^2 - 10y + 6) \div 2y$

2. $(-9x^2 + 15x - 12) \div 3x$

3. $(12t^2 - 16t + 8) \div 4t$

4. $(-20w^2 + 12w + 8) \div 2w$

Flashcard App

 **Probability**

MUST KNOW

⚡ Probability assigns a numerical value to the likelihood of a particular outcome relative to all possible outcomes.

⚡ A combination is a selection of items from a larger set in which the order of the items does *not* matter.

⚡ A permutation is an arrangement of items in which the order of items *does* matter.

⚡ A sample space is the set of all possible outcomes in a probability experiment.

⚡ The counting principle states that if event *A* can happen in *m* ways and event *B* in *n* ways, then events *A* and *B* can happen in *m* × *n* ways.

any people think of probability as some complicated mathematical process by which we make predictions about the future. While probability is about the likelihood of future outcomes, it's really not all that complicated. The key to understanding probability is the ability to identify precisely all the possible outcomes a situation presents and the favorable outcome(s). The **favorable outcome** is the outcome that we're looking for.

Probability is the branch of mathematics that assigns a numerical value to the likelihood of a favorable outcome among all possible outcomes. This is **classical probability theory**, which is based on the assumption that all possible outcomes are equally likely and can be identified.

Probabilities range from 0 to 1, where 0 means the likelihood of an event is *impossible* and 1 means it is *certain* to occur. Most events fall somewhere in between these two extremes. If we think of probability as a line segment labeled 0 to 1, it would look like this:

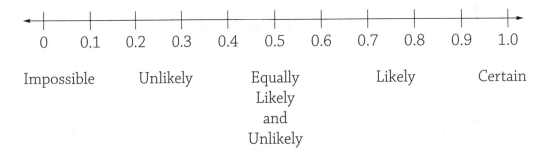

Counting Outcomes

The set of all possible outcomes in a probability problem is called the **sample space**. When flipping a coin, the sample space is small since there are only two possible outcomes: heads or tails. When tossing a standard six-sided die, the sample space is somewhat larger, but it's still easy to count: 1, 2, 3, 4, 5, and 6. Not all probability situations are so easy, and counting all possible outcomes can be trickier than it might seem. One way to be sure that we haven't missed a possible outcome is to organize them all using a **tree diagram** that shows each outcome on a separate branch.

▶ A school T-shirt comes in three sizes (small, medium, and large) and three colors (red, white, and blue). How many T-shirts of different sizes and colors are there? Show your answer in a tree diagram.

▶ To make a tree diagram, first list all the sizes. Then, add branches that show the different colors.

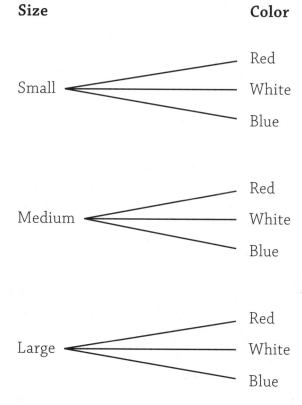

Size	Color
Small	Red / White / Blue
Medium	Red / White / Blue
Large	Red / White / Blue

▶ Count the number of outcomes.

▶ Based on size and color, there are 9 different school T-shirts.

Suppose your school T-shirts came in 5 sizes and 10 colors. Clearly, making a tree diagram would be more complicated than our preceding example. Fortunately, there's an easier way to determine how many possible outcomes there are in this situation.

According to the **counting principle**, if event A can happen in m ways and event B can happen in n ways, then events A and B can happen in $m \times n$ ways. In our example, if there are 5 sizes and 10 colors of T-shirts, we can easily calculate that there are $m \times n$ (5 × 10), or 50, outcomes.

You're very likely familiar with combination locks that are used on lockers in schools and gyms. Suppose the numbers for opening a lock are 6, 11, and 23. All possible arrangements of these numbers include:

6, 11, 23	11, 6, 23	23, 11, 6
6, 23, 11	11, 23, 6	23, 6, 11

Now, the fact is that only one arrangement of these three numbers will actually open the lock. That's because your lock is not a combination lock. If it were, then any of these sequences of numbers would open it. Actually, your lock is a permutation lock: you must use not only the correct numbers, but you also must enter them in the correct order. A **permutation** is an arrangement of items, or outcomes, in which order *is* important.

Making a tree diagram or list of outcomes can be easy, as long as we are dealing with a few items. If the number of items is large, however, we can spend a lot of time identifying and writing each outcome, and there's always the likelihood that we might leave one out. There is a simpler method that is less prone to errors: using factorials. A **factorial** is the product of an integer and all the integers less than it. A factorial is written as the integer followed by an exclamation point. So 4! is: $4 \times 3 \times 2 \times 1 = 24$.

The following example shows how to use factorials to find the number of permutations in a given situation.

EXAMPLE

▶ Touseen plans to watch 6 new movies during her two-week spring break. In how many different orders can she watch the movies?

▶ Write a factorial to represent the number of permutations. Since there are 6 movies, we write: 6!

▶ Next, calculate the product of the integer and those below it: $6! = 6 \times 5 \times 4 \times 3 \times 2 \times 1 = 720$.

▶ There are 720 possible arrangements, or permutations, for Touseen to watch the movies.

In some problems like the one that follows, you only need to deal with a limited number of choices.

EXAMPLE

▶ Noah wants to arrange three model cars on a shelf. He has 8 cars from which to choose. How many different arrangements of three cars can he make?

▶ Since Noah is choosing 3 cars, write the number of choices he has for each of the three spots: 8, then 7, and then 6. Notice that the numbers decrease because after Noah chooses the first car from the group of 8, there are only 7 cars left, and when he chooses another, there are only 6 cars left for the final choice.

▶ Multiply the number of choices for each of the three spots Noah can choose cars for.

$$8 \times 7 \times 6 = 336$$

▶ There are 336 permutations, or possible arrangements.

In some cases, when dealing with a group of items, the order in which they appear is not important, but avoiding duplication is essential. For example, if we flip a coin as heads and spin a 6 on a spinner, it doesn't really matter which event occurs first. Flipping a heads and then spinning a 6 is the same as spinning a 6 and then flipping a heads. A **combination** is a group of items, or outcomes, in which order does *not* matter.

Let's look at an example that shows why order does not matter when dealing with combinations.

EXAMPLE

▶ The soccer team is looking to select 2 new players for the school team. There are three qualified players: Adam, Bess, and Charles. The team captain writes the name of the three players on separate pieces of paper, places them in a bag, mixes them up, and selects two slips at random. The first thing to do is make a table that shows all possible combinations of pairs of players that might occur.

Choice A Player 1	Choice B Player 2
Adam	Bess
Adam	Charles
Bess	Adam
Bess	Charles
Charles	Adam
Charles	Bess

▶ Note that choosing Adam and Bess is the same as choosing Bess and Adam. That's exactly why it's important to eliminate all overlaps when dealing with combinations! Cross out the duplicates.

Choice A Player 1	Choice B Player 2
Adam	Bess
Adam	Charles
~~Bess~~	~~Adam~~
Bess	Charles
~~Charles~~	~~Adam~~
~~Charles~~	~~Bess~~

▶ There are three combinations involved in the selection of two players.

We can eliminate duplication when finding combinations by dividing the number of choices for each spot by the number of ways to arrange those choices. The following example shows how to do this.

EXAMPLE

▶ Bridget wants to put three seashells on a shelf. She has 10 seashells from which to choose. How many different groups of 3 seashells can Bridget make?

▶ First, notice how this example is different from the last permutation example. In the preceding permutation problem, Noah is *arranging* cars. In this problem, Bridget is simply *grouping* seashells. In other words, the order she puts the seashells in does *not* matter. A group consisting of seashells *1, 2, 3* is the same as group containing seashells *3, 2,* and *1*. This difference between arranging and grouping is what makes this is a combination problem.

▶ Since Bridget is choosing 3 seashells, find the product of the number of choices she has for the three spots.

$$10 \times 9 \times 8 = 720$$

> Then, divide the number of choices by the number of ways to arrange the 3 seashells, that is, by 3!

$$\frac{10 \times 9 \times 8}{3 \times 2 \times 1} = \frac{720}{6} = 120$$

> There are 120 combinations or ways that Bridget can arrange 3 seashells from her collection of 10 seashells.

Simple Probabilities

We can calculate the probability of a simple independent event such as tossing a standard number cube by using the formula:

$$\text{Probability of an event }(P) = \frac{\text{Number of favorable outcomes}}{\text{Number of possible outcomes}}$$

When we toss a die, there are 6 possible outcomes: 1, 2, 3, 4, 5, and 6. The probability of the outcome we choose (the favorable event) depends on what we choose. If we choose just one number—say, 3—then the probability is:

$$P(3) = \frac{1}{6}$$

The probability of choosing 1, 2, 4, 5, or 6 is the same: $\frac{1}{6}$. Note that we can also write this probability as a decimal or as a percent: $\frac{1}{6}$ approximately equals $0.1\overline{66}$, or 16.66%.

Let's try a few more examples to get a good feel for calculating simple probabilities.

EXAMPLE

▶ If you toss a standard die, what is the probability that you will roll a number greater than 3? Write the probability as a fraction.

▶ There are six possible outcomes (1, 2, 3, 4, 5, 6), and three outcomes are favorable (4, 5, and 6).

$$P(4, 5, \text{ or } 6) = \frac{3}{6} = \frac{1}{2}$$

▶ The probability of rolling a number greater than 3 on a number cube is $\frac{1}{2}$.

When solving probability problems, it's important to think carefully about what is being asked.

EXAMPLE

▶ If we roll a die, what is the probability that the number we roll is irrational?

▶ There are 6 possible outcomes (1, 2, 3, 4, 5, 6), but none is favorable since all the numbers on a standard number cube are integers and, therefore, rational.

$$P(\text{irrational number}) = \frac{0}{6} = 0$$

▶ The probability of rolling an irrational number is 0. This means that the occurrence of the favorable outcome is *impossible*.

Let's consider an event that is certain to happen.

EXAMPLE

▸ If you roll a die, what is the probability that the number you roll is rational?

▸ There are 6 possible outcomes (1, 2, 3, 4, 5, 6). Each number is an integer; therefore, all outcomes are favorable.

$$P(\text{rational number}) = \frac{6}{6} = 1$$

▸ The probability of rolling an integer is 1. This means that a favorable outcome is *certain* to occur.

Some problems may ask us to find the probability of an event that will *not* happen. It's sometimes easier to find the answer by calculating the complement of the event. The **complement** of an event is the set of all outcomes that are *not* favorable. Taken together, the favorable event and its complement comprise all possible outcomes.

EXAMPLE

▸ Suppose a bag contains 15 chips: 3 red, 3 blue, 3 green, 3 black, and 3 white. What is the probability that, without looking, you will select a chip that is not white? Write the probability as a fraction, a decimal, and a percent.

▸ Find the probability of selecting a white chip.

$$P(\text{white}) = \frac{3}{15} = \frac{1}{5}$$

▸ The probability of all possible events equals 1. The complement of $\frac{1}{5}$ represents all events that do not

BTW

The "odds" that an event will occur has a different meaning from the "probability" of an event occurring. Odds are the ratio of the likelihood that the event will occur divided by the likelihood that the event won't occur. The probability that we will roll 4 on a number cube is $\frac{1}{6}$. However, the odds of rolling 4 are 1 in 5.

involve selecting a white chip. So, we subtract $\frac{1}{5}$ from 1.

$$P(\text{not white}) = 1 - \frac{1}{5} = \frac{4}{5}$$

▶ The probability that you will *not* select a white chip is $\frac{4}{5}$, or 0.8 (80%).

Compound Events

The probability of a **compound event** involves finding the probability of two or more simple events occurring. There are actually two different types of compound events. In one situation, we may be asked to find the probability of event *A* occurring *and* event *B* occurring. The other type of compound event asks us to find the probability of event *A* occurring *or* event *B* occurring.

Let's take a look at the first type of probability involving compound events—the likelihood that event *A and* event *B* will occur.

▶ You toss two quarters at the same time. What is the probability that both coins will land heads up? Write the probability as a fraction, a decimal, and a percent.

▶ To find the probability that both coins will land heads up, use the formula $P(A \text{ and } B) = P(A) \times P(B)$.

$$P(\text{both heads}) = \frac{1}{2} \times \frac{1}{2} = \frac{1}{4}$$

▶ Find the decimal and percent value of the fraction.

$$\frac{1}{4} = 0.25$$

$$\frac{1}{4} = 25\%$$

▶ The probability of tossing two quarters that both land heads up is $\frac{1}{4}$, or 0.25 (25%).

Independent Events

Since the outcome of each coin toss in the preceding example does not affect the probability of the outcome of another coin toss, compound events like this one are called **independent events**. From our example, we can generalize a formula to use when calculating the probability of compound events that are independent:

$$P(A \text{ and } B) = P(A) \times P(B)$$

When a problem involving compound events uses the word *or*, we must calculate the probability of each event and then find the sum of their probabilities:

$$P(A \text{ or } B) = P(A) + P(B)$$

Here's an example that shows how to calculate the probability of two events when they have no outcomes in common.

EXAMPLE

▶ A card with a spinner has 8 equal sections numbered 1 through 8. What is the probability that you will spin an odd number less than or equal to 3 or a number greater than 5? Write the probability as a fraction, a decimal, and a percent.

▶ Calculate the probability of spinning an odd number less than or equal to 3. There are two favorable outcomes: 1 and 3.

$$P(\text{odd } n \leq 3) = \frac{2}{8}$$

▶ Calculate the probability of spinning a number greater than 5. There are three favorable outcomes: 6, 7, and 8.

$$P(n > 5) = \frac{3}{8}$$

▶ Determine the probability that one or the other event will occur by finding the sum of the probabilities.

$$\frac{2}{8} + \frac{3}{8} = \frac{5}{8}$$

▶ Find the decimal and percent values of the fraction.

$$\frac{5}{8} \approx 0.625$$

$$\frac{5}{8} \approx 62.5\%$$

▶ The probability of spinning an odd number less than or equal to 3 *or* a number greater than 5 is $\frac{5}{8}$, or approximately 0.625 (62.5)%.

In the preceding example, there are no overlaps between the events. In other words, the events are mutually exclusive, or **disjointed**, meaning they have no outcomes in common. Sometimes, two events **overlap**; that is, they share one or more outcomes. When this happens, it's important to avoid counting the overlap twice, so we use this formula:

$$P(A \text{ or } B) = P(A) + P(B) - P(A \text{ and } B)$$

The following example shows how to calculate the probability of independent events when there is one or more overlapping outcomes.

EXAMPLE

▶ You select a card at random from a standard deck of cards. What is the probability that you will choose a queen *or* a diamond? Write the probability as a fraction, a decimal, and a percent.

▶ There are 52 cards in a standard deck of cards and 4 of them are queens, so the probability of selecting a queen is:

$$P(\text{queen}) = \frac{4}{52}$$

▶ There are 52 cards in a standard deck of cards and 13 of them are diamonds. Therefore, the probability of selecting a diamond is:

$$P(\text{diamond}) = \frac{13}{52}$$

▶ Since one card is both a queen and a diamond, we must subtract $\frac{1}{52}$ when the calculating the probability of drawing a queen *or* a diamond.

$$P(\text{queen or diamond}) = \frac{4}{52} + \frac{13}{52} - \frac{1}{52} = \frac{16}{52} = \frac{4}{13}$$

▶ Find the decimal and percent value of the fraction.

$$\frac{4}{13} \approx 0.3077 \rightarrow 0.3077 \approx 30.77\%$$

▶ The probability of selecting a card that is a queen or a diamond from a standard deck is $\dfrac{4}{13}$, or approximately 0.3077 or 30.77%.

Dependent Events

Some compound probability problems involve events in which one outcome *does* change the probability of the other outcome. Compound events like these are called **dependent events**. In these situations, we use the formula that follows.

$$P(A \text{ and } B) = P(A) \times P(B \text{ given } A)$$

Here's an example of the steps involved in calculating the probability of compound dependent events.

EXAMPLE

▶ We have a bag with five red marbles and five blue marbles. We want to select two marbles to give a friend. We chose the first marble and, without replacing it into the bag, we chose the second marble. What's the probability that we will randomly select two red marbles? Write the probability as a fraction, a decimal, and a percent.

▶ Calculate the probability of selecting a red marble on the first draw.

$$P(\text{red}_1) = \frac{5}{10} = \frac{1}{2}$$

▶ Calculate the probability of selecting a red marble on the second draw. Remember that the first marble was not put back into the bag, so this event is dependent on the first event.

$$P(\text{red}_2) = \frac{4}{9}$$

▶ Evaluate the probability that both marbles we choose will be red.

$$P(\text{red}_1 \text{ and } \text{red}_2) = \frac{1}{2} \times \frac{4}{9} = \frac{4}{18} = \frac{2}{9}$$

▶ Find the decimal and percent values of the fraction.

$$\frac{2}{9} \approx 0.\overline{22}$$

$$\frac{2}{9} \approx 22\%$$

▶ The probability of selecting two red marbles on the first and second tries is $\frac{2}{9}$, or 0.22, or 22%.

Here's another problem that involves the probability of two dependent events.

▶ A cookie jar contains 9 chocolate chip, 3 pecan sandy, and 3 oatmeal cookies. You select one cookie at random and eat it. Then, you select a second cookie at random and eat it. What is the probability that both cookies you selected were chocolate chip cookies? Write the probability as a fraction.

▶ Calculate the probability that the first cookie you select is a chocolate chip cookie:

$$P = \frac{9}{15} = \frac{3}{5}$$

▸ Calculate the probability that the second cookie you select is a chocolate chip cookie. Remember that you ate the first cookie!

$$P = \frac{8}{14} = \frac{4}{7}$$

▸ Multiply the probabilities of the two dependent events.

$$\frac{3}{5} \times \frac{4}{7} = \frac{12}{35}$$

▸ Find the decimal and percent value of the fraction.

$$\frac{12}{35} \approx 0.3428 \rightarrow 0.3428 = 34.28\%$$

▸ The probability of selecting two chocolate chip cookies from the cookie jar is $\frac{12}{35}$, or approximately 0.3428 or 34.28%.

Experimental Probability

As we know, theoretical probability is based on what is expected to happen given all possible outcomes and all favorable outcomes. With theoretical probability, we just need a few facts and the ability to analyze a situation. **Experimental probability**, in turn, is based on the outcomes observed in real-world trials. In other words, we perform "experiments" and collect data about the outcomes that occur. In both types of probability—theoretical and experimental—we are still using the same fundamental principle:

$$\text{Probability of an event } (P) = \frac{\text{Number of favorable outcomes}}{\text{Number of possible outcomes}}$$

In other words, it's just the methods by which we determine the number of outcomes that are different.

Here's a good way to think of how these two types of probability differ. We know that the probability of tossing a 3 on a six-sided die is $\frac{1}{6}$. So, based on theoretical probability, if we tossed a die 60 times, we would expect to toss 3 exactly ten times. However, what happens if we actually roll a die 60 times? Well, there's no guarantee that we will toss a 3 exactly ten times. We may toss a 3 only 6 times, 7 times, or 11 times. In fact, it's possible, though extremely unlikely, that we may toss 3 every time we roll the number cube!

In many real-world situations, it is virtually impossible to calculate the theoretical probability of an event. This fact is what makes experimental probability so useful. For example, suppose a company that manufactures cell phones examines 100 phones that just rolled off its production line and finds 12 of the phones defective in some way: $\frac{12}{100} = 0.12 = 12\%$.

Using this result, the company can easily predict that of the next 1,000 cell phones it makes, about 120 of them will be defective, or that of the next 5,000 about 600 will have defects.

EXAMPLE

▶ The results of a survey of 50 students at a high school show the colors of students' shirts or blouses on a given day. What is the probability that the next student surveyed has a green shirt or blouse?

Shirt Color	Number of Students
White	22
Blue	20
Green	6
Orange	2

▶ Find the ratio of students wearing green shirts or blouses to all students surveyed. Use the data to calculate the probability of wearing green.

$$P(\text{green}) = \frac{6}{50} = 0.12 = 12\%$$

▶ There's a 12% chance the next student surveyed will be wearing a green shirt or blouse.

Geometric Probability

The basic types of probability that we have dealt with so far involve "discrete" situations such as identifying all outcomes and all favorable outcomes. Geometric probability is a tool that allows us to calculate the likelihood of an event in a "continuous" situation. For example, geometric probability involves finding the likelihood that a point falls on a line segment or in a certain part of a plane figure.

▶ Line segment $\overline{CE}$ is on line segment $\overline{AB}$. What is the probability that a point chosen at random on $\overline{AB}$ falls on $\overline{CE}$?

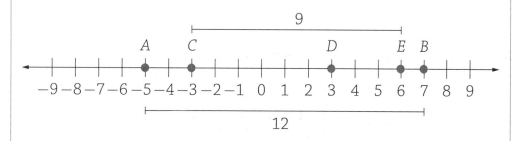

▶ Find the length of line segment $\overline{AB}$ and the length of line segment $\overline{CE}$.

Length of $\overline{AB} = 12$

Length of $\overline{CE} = 9$

▶ Find the ratio of the length of line segment $\overline{CE}$ to the length of line segment $\overline{AB}$.

$$\frac{\text{Length of } \overline{CE}}{\text{Length of } \overline{AB}} = \frac{9}{12} = \frac{3}{4}$$

▶ The probability that a point chosen at random on $\overline{AB}$ falls on $\overline{CE}$ is $\frac{3}{4}$, or 0.75, or 75%.

Let's try another example.

EXAMPLE

▶ If a dart is thrown at the square game board shown below, what is the probability that it will land in the shaded area?

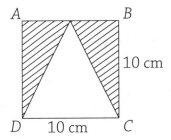

▶ Find the area of the square game board and the area of the triangle.

$$A(\text{game board}) = s^2 = 10 \times 10 = 100 \text{ square centimeters}$$

$$A(\text{triangle}) = \frac{1}{2}bh = \frac{1}{2}(10 \times 10) = \frac{1}{2}(100)$$

$$= 50 \text{ square centimeters}$$

▶ Find the ratio of the area of triangle to the area of the square.

$$\frac{\text{Area of triangle}}{\text{Area of game board}} = \frac{50}{100} = \frac{1}{2}$$

▶ Subtract the area of the triangle from the area of the game board.

$$1 - \frac{1}{2} = \frac{1}{2}$$

▶ The probability that a dart will land on the shaded portion of the game board is $\frac{1}{2}$, or 0.50, or 50%.

 IRL Geometric probability plays an important role in space exploration, the assessment of storm damages, the search for oil and natural gas, and the efficiency of manufacturing processes.

EXERCISES

EXERCISE 11–1

Identify the permutations for each situation.

1. You have 4 science documentaries you want to watch. How many different orders can you watch the science documentaries in?

2. There are 6 soccer teams in the finals of a tournament. In how many permutations can the teams place first, second, and third?

EXERCISE 11–2

Find all the combinations possible for each problem.

1. A local taco stand sells 5 different kinds of tacos and 2 different drinks. If you buy 1 taco and 1 drink, how many possible combinations did you have to choose from? Make a tree diagram to explain your answer.

2. Three balls are labeled 1, 2, and 3. How many different ways can the balls be arranged? Make a table to explain your answer.

EXERCISE 11–3

Find the simple probability for each event.

1. A spinner has 8 equal sections numbered 1 through 8. What is the probability that you will spin a number greater than 6? Write the probability as a fraction.

2. A bag contains 5 red chips, 10 white chips, 5 blue chips, and 10 black chips. If you select a chip at random, what is the probability that it will be blue? Write the probability as a decimal.

3. Five tiles are placed in a row to spell the word SMILE. The tiles are placed in a paper bag. If you select one tile from the bag, what is the probability it will be a vowel? Write the probability as a percent.

4. The numbers 20 through 30 are written on separate cards and placed in a bag. What is the probability that a randomly selected card is an even number? Write the probability as a fraction.

EXERCISE 11–4

Find the probability for each of the dependent or independent events described.

1. You flip a coin and roll a standard number cube. What is the probability that the coin will land heads up and the number rolled is less than 3? Write the probability as a fraction.

2. There are ten tiles numbered 1–10 placed face down. You select two tiles at random. What is the probability that both tiles will be greater than 5? Write the probability as a decimal to the nearest hundredths place.

3. A bag holds 10 counters: 4 are blue, 3 are red, and 3 are green. If Danielle randomly selects two counters from the bag without replacing the first, what is the probability that both counters will be blue? Write the probability as a fraction.

4. A bag contains 12 Macintosh apples and 8 Fuji apples. Sean randomly takes an apple from the bag and does not replace it. He then takes a second apple from the bag. What is probability that both are Fuji apples? Write the probability as a decimal to the nearest hundredths place.

EXERCISE 11–5

Find the experimental probability for each situation.

1. In a random survey, 200 students at Pembroke High School were asked to choose their favorite movie genre. The table below shows the results of the survey. Based on the survey, what is the probability that the next randomly selected student will say that comedy is his or her favorite movie genre? Write the probability as a decimal to the nearest hundredths place.

Favorite Movie Genre	Number of Students
Science Fiction	55
Comedy	40
Action	47
Mystery	35
Drama/Romance	23

2. A spinner is divided into six equal sections numbered from 1 to 6. Jeni spins 30 times. The table summarizes the results of her trials. What is the experimental probability that the next time Jeni spins, the spinner will land on 1? Write the probability as a percent.

Results	1	2	3	4	5	6
Number of Spins	6	5	4	5	6	4

EXERCISE 11–6

Use the figures to find the geometric probability of each event.

1. Line segment $\overline{JL}$ is on line segment $\overline{JK}$. What is the probability that a point chosen at random on $\overline{JK}$ falls on $\overline{JL}$?

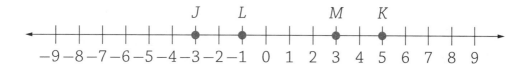

2. If a dart is thrown at the square game board shown below, what is the probability that it will land in the shaded area? Write the probability as a percent to the nearest tenths place.

4 cm

Data and Statistics

MUST ⚡ KNOW

⚡ Statistics is the study of methods for collecting, organizing, and analyzing numerical data.

⚡ Mean, median, and mode are measures of central tendency used to summarize large collections of data.

⚡ Bar graphs use bars of different lengths to represent discrete data, whereas histograms use bars to represent continuous data.

⚡ Circle graphs present data as proportional sections of a circle, whereas line graphs use a line to show changes in data over time.

⚡ Stem-and-leaf plots present data in a table in which the first digits are listed as stems and the last digits are represented as leaves. Box-and-whisker plots use a number line and a box to provide a visual representation of the median, quartiles, and extremes of a data set.

⚡ Scatter plots use points to display the relationship between two sets of data.

tatistics are impossible to avoid in modern life. In fact, many people seek out data and thrive on it. They reference it in discussions with family members, neighbors, and coworkers. When was the last time you got into a conversation with someone about how well (or poorly) a particular sports team or player is doing? If you're like most fans, you probably referred to some important numbers to support your opinion.

Data and statistics often play more than a "fun" role in our lives. A smart consumer about to buy a car or a house compares interest rates on loans to see which will save money. A small business owner usually looks at data to compare sales or profits over time. And, of course, doctors, physical therapists, and other health professionals rely on data to decide which drugs or treatment procedures are best.

Over the course of the next two days, make a list of all the times you come in contact with data and statistics. You're likely to be surprised by how frequently you "bump" into such numbers, not just in math class but in your day-to-day life!

Measures of Central Tendency

There's a lot that can be known by analyzing and evaluating a collection of numbers, or **data**. **Measures of central tendency** are statistical values that are used to summarize a set of data by helping to identify its center, or middle values. The three most common measures of central tendency are the mean, median, and mode. Let's explore what these measures are and what they tell us about a set of data.

The data set below shows the number of home runs in the past season by the top eleven players in a state's high school baseball league. Notice that the elements in this data set are arranged in order from greatest to least.

21, 18, 16, 15, 15, 13, 12, 11, 9, 8, 5

Based on the data, the **mean**, or average number of home runs for the season is 13. To find the mean we first need to find the sum of the home runs. Then, we must divide the sum by the number of items in the data set. If we add all the home run scores, the sum is 143 points. To find the mean, we simply divide the sum of 143 by 11 since this is the number of home run scores, or items, in the data set: $143 \div 11 = 13$.

When the items in a data set are arranged from least to greatest or greatest to least, it's important to look at the middle number(s) called the **median**. Since this data set has 11 items, the sixth number—13—is the median.

21, 18, 16, 15, 15, **13**, 12, 11, 9, 8, 5

The number that occurs most frequently in a data set is called the **mode**. In this array, the number 15 appears twice, and all the other scores appear just once. Therefore, the mode is 15.

Sometimes, it useful to look at the **range**, or the difference between the greatest value and the least value in a data set. In the data set of home runs scored last season, the range is 16, since 21 minus 5 equals 16.

EXAMPLE

▶ Thirteen students in Mr. Crenshaw's science class took their final exams. Their scores are shown in the data set below. What is the average or mean test score on the final exam?

55, 60, 63, 70, 72, 78, 80, 85, 85, 88, 91, 93, 94

▶ The sum of the 13 scores is 1,014. To find the mean, divide the sum by the number of items, which is 13: $1,014 \div 13 = 78$.

▶ The mean score on the final science test is 78.

Suppose there are 12 test scores instead of 13. How do you determine the median when there are two middle numbers?

▶ The test scores on the final exam of the 12 students in Mr. Cranshaw's science class are shown below. What is the median score?

▶ First, make sure the scores are in order from least to greatest. Identify the middle numbers.

60, 63, 70, 72, 78, **80**, **85**, 85, 88, 91, 93, 94

▶ Then, find the average of the two middle numbers.

$80 + 85 = 165$

$165 \div 2 = 82.5$

▶ The median score on Mr. Cranshaw's final science exam is 82.5.

Identifying the mode of data set is very easy. All you have to do is look for the number that occurs most frequently.

▶ The data set below shows the scores of students in Ms. Peterson's class on their language arts midterm exam. What is the mode of this data set?

▶ Identify the score that appears most frequently.

62, 69, 75, 79, 81, 83, 83, 87, 90, 93, 94

▶ Notice that there are two scores of 83.

▶ The mode of this data set is 83.

As noted previously, all we have to do to find the range of a data set is subtract the least value from the greatest value. That's exactly why it so important to make sure the numbers are ordered from the least value to greatest.

EXAMPLE

▶ The data set that follows shows the prices of different brands of blue jeans at a local department store. What is the range of this data set?

$43.95, $59.95, $35.00, $79.50, $125.00, $98.00

▶ Arrange the prices in order.

$35.00, $43.95, $59.95, $79.50, $98.00, **$125.00**

▶ Since 125 minus 35 equals 90, the range of this data set is 90.

Bar Graphs and Histograms

A common way to compare quantities is to use vertical or horizontal bars of different lengths to represent data in a **bar graph**. For example, the bar graph that follows shows the most popular types of pets owned by students in a survey.

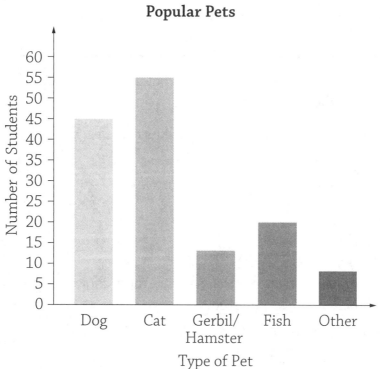

Popular Pets

Most bar graphs make comparisons using vertical bars. When reading a vertical bar graph, look at the top of edge of each bar. Then locate the value that's listed at the left along the vertical axis.

▶ Based on the bar graph on the previous page, which pet is the most popular and how many students chose it?

▶ Look at the height of the bars to find the tallest one and read the label below it on the horizontal axis. The tallest bar represents cats.

▶ Look at the top edge of the bar representing cats. Then, locate the number on the vertical axis that corresponds to the top of the bar. 55 students selected cats as their favorite pet.

▶ Cats are the most popular pets, having been chosen by 55 of the students surveyed.

We can also use bar graphs to do some limited calculations.

▶ Based on the Popular Pets bar graph, what fraction of the students chose cats as their top choice?

▶ Read the graph to find how many students chose each type of pet.

Dog = 45
Cat = 55
Gerbil/Hamster = 13
Fish = 20
Other = 7

▶ Add the number of students that selected these choices to find the total number of people surveyed.

$$45 + 55 + 13 + 20 + 7 = 140$$

▶ Since 55 respondents chose cats as their favorite pet, the fraction of students that chose cats is $\dfrac{55}{140}$, or $\dfrac{11}{28}$.

With horizontal bar graphs, the scale of values is along the bottom of the graph. Therefore, we read down from the right edge of the bar to the values along the horizontal axis. In the bar graph that follows, the favorite colors chosen by students in a survey appear along the vertical axis.

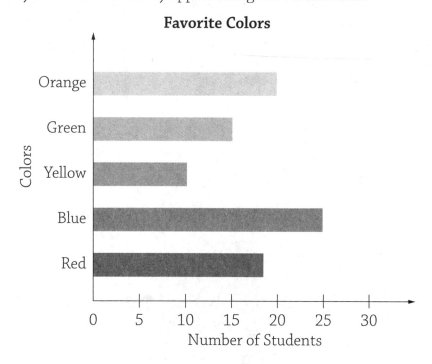

Favorite Colors

EXAMPLE

▶ Based on the preceding bar graph, which color was selected by the most students?

▶ Identify the bar that is longest and read the label at the left along the horizontal axis. The longest bar represents the number of students that chose blue as their favorite color.

▶ Look at the right edge of the bar that represents blue and then look at its value, presented along the horizontal axis. 25 students selected blue as their favorite color.

▶ Blue was selected as the favorite color by the largest number of students surveyed, 25 in all.

Histograms are displays of data similar to bar graphs, but there are no gaps between the bars. That's because histograms represent ranges of continuous data with no gaps. The bars in bar graphs represent discrete, or separate, data. The histogram below shows the number of runs a high school baseball team scored in games during the last year's season.

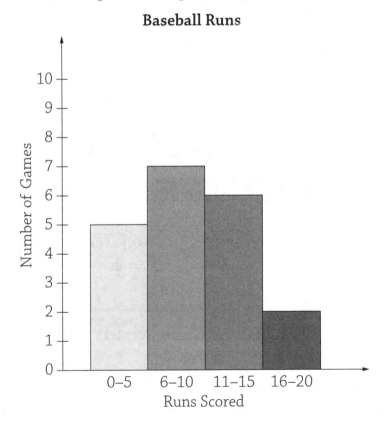

Baseball Runs

▶ Look at the preceding histogram, which shows runs scored in baseball games last season. In which range were the most number of runs scored? In how many games did the number of runs scored fall within this range?

▶ Look at the height of the bars to identify the one that is tallest. Then, read the label below it along the horizontal axis that tells the range of runs this correpsonds to—6 to 10 games.

▶ Find the top edge of the tallest bar. Then, locate the number of games played that had scores within this range on the vertical axis. According to the data display, there were 7 such games.

▶ The most number of runs scored in games last season were in the 6–10 range. Scores within this range were achieved in 7 games.

Circle Graphs and Line Graphs

A **circle graph**, or a pie chart, is a graph that displays data by dividing a circle into sections, each of which represents a quantity related to a specific category. The entire circle represents 100% of the data. The larger the area of a section, the greater the quantity it represents in relation to the whole. Let's look at the pie chart below, which shows student choices for their favorite type of music.

Favorite Types of Music

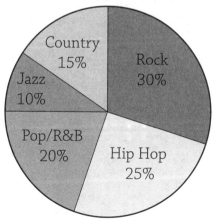

Just by looking at the circle graph, it is easy to see that the section labeled Rock is the largest. This means rock is the students' favorite type of music. Notice that the label for this section also indicates 30%, which is the greatest percent listed in the circle graph. Most, but not all, circle graphs give the actual percents of each category.

EXAMPLE

▶ Look at the preceding graph showing favorite types of music. If 100 students were surveyed and one student is chosen at random, what is the probability that the student favors pop/R&B or hip hop? Write the probability as a percent.

▶ To find probability, find the number of favorable outcomes by adding the number of students who selected each of the two categories.

Pop/R&B = 20%, or 20 students.
Hip Hop = 25%, or 25 students.
$20 + 25 = 45$

▶ Write a fraction that shows the sum of the two categories as a numerator over a denominator of 100.

$$\frac{45}{100}$$

▶ Change the fraction into a decimal and a percent.

$$\frac{45}{100} = 0.45 = 45\%$$

▶ The probability that a student selected at random will choose pop/R&B or hip hop as their favorite type of music is 45%.

Circle graphs also allow us to make comparisons.

EXAMPLE

▶ Refer again to the music graph. Which type of music was chosen by half as many students as those who chose rock?

▶ Look at the circle and find the section labeled Rock. It represents 30% of the students. Now, find the section that is half as large as the Rock section. Country represents 15% of the students.

▶ Half as many students chose country as chose rock.

A **line graph** shows changes in data over time by using points connected by line segments. When a line graph is created, points representing specific data are plotted first and then they are connected in sequence. In this way, a line graph makes **trends**, or changes in the data, easy to see. The line graph below shows the number of graphic novels a bookstore sold in 2019.

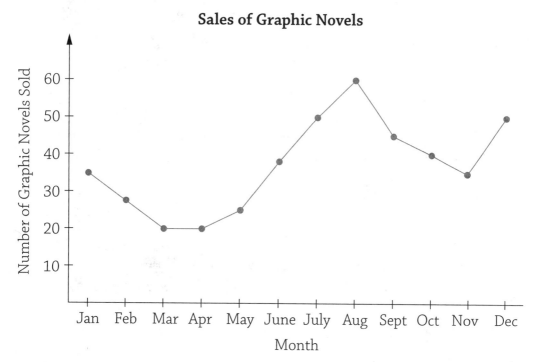

Sales of Graphic Novels

EXAMPLE

▶ Look at the line graph on the previous page. During which months were sales of graphic novels the lowest? How many graphic novels were sold in these months?

▶ Look for the points where the line is at its lowest. Read the horizontal axis to identify the month(s) to which these points correspond. March and April are the lowest points on the line graph.

▶ Look at the vertical axis to find the number of sales for these months. In March and again in April, 20 graphic novels were sold.

▶ March and April had the lowest sales, with 20 graphic novels sold in each month.

Line graphs also help us see trends in data.

EXAMPLE

▶ Based on the Graphic Novels line graph, during which four-month period did the sales of graphic novels increase the most? What was the range of the number of graphic novels sold per month during this four-month period?

▶ Look for the four consecutive points where the line rises the most. Read the horizontal axis to identify the months to which these points correspond. Sales rose the most during May, June, July, and August.

▶ Look at the vertical axis to find the least number of graphic novels sold during this period. Then, identify the most number of graphic novels sold in a month during the same four months.

The least number of graphic novels sold is 25 during May.
The most number of graphic novels sold is 60 during August.

▶ Find the range by subtracting the least value from the greatest value.

$$60 - 25 = 35$$

▶ The sales of graphic novels increased the most during May, June, July, and August. The range of the number of graphic novels sold during this four-month period is 35.

Stem-and-Leaf Plots

A **stem-and-leaf plot** is a compact way of displaying and comparing a large amount of data. Stem-and-leaf plots show data in an organized table in which the first digit(s) are presented as "stems" and the last digits are presented as "leaves." This type of data display is helpful when we want to see how data are distributed. Let's see how a stem-and-leaf plot works.

EXAMPLE

▶ The stem-and-leaf plot below shows a high school basketball team's scores during a recent season.

Basketball Scores

Stem	Leaf
5	2 4
4	3 5 8
3	2 3 5 7 9
2	1 2 6 6 6 8
1	1 9
0	4 8

3 | 2 = 32 points

▶ What were the team's lowest and highest scores?

EXAMPLE

▶ To find the lowest score, look for the stem with the lowest number and then find the leaf with lowest number. The lowest score is 4.

▶ To find the highest score, find the stem with the greatest number and then find the leaf with the greatest number. The highest score is 54.

▶ The team's lowest score was 4 and its highest score was 54.

Let's look at how to find the mode in a stem-and-leaf plot.

▶ Refer again to the Basketball Scores stem-and-leaf plot. What was the team's most frequent score?

▶ Scan the leaf data for each stem. Look for the number that occurs most frequently in all the stems. In stem 2, three 6s appear in the leaf. No other stem has more than 2 leafs.

▶ The mode is 26, since it is the team's most frequent score.

Box-and-Whisker Plots

A **box-and-whisker plot** uses three measures to represent a data set as four distinct parts. The data items in this type of display are organized from least to greatest:

■ The median is used to divide the box-and-whisker plot into a **lower half** and **an upper half**. Each half is divided into two quartiles.

■ The **lower quartile** is the median of the lower half of the data.

■ The **upper quartile** is the median of the upper half of the data.

A box-and-whisker plot shows the distribution of items, or how spread out the data are. Let's say that Joy scored 62, 65, 74, 75, 78, 82, and 94 on her science quizzes. Here is a box-and-whisker plot that shows Joy's quiz scores.

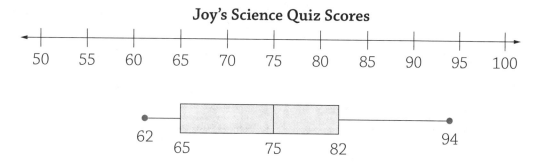

Joy's Science Quiz Scores

A quick glance at this box-and-whisker plot reveals key characteristics of this data set. The box, called the **interquartile range**, represents half the data that comprise the middle values of set. The whisker, or line, to the left of the box is called the **lower quartile**. It represents one-quarter of the data with the lowest values. Likewise, the whisker to the right of the box shows the data in the **upper quartile**. The two endpoints represent the extremes—the lowest value and the greatest value—of the data set.

EXAMPLE

▸ Refer to the box-and-whisker plot Joy's Science Quiz Scores above. What is the median of her quiz scores?

▸ Look at the box section of the data display. Find the line within the box that represents the median, or middle, quiz score.

▸ The median of Joy's scores on her science quizzes is 75.

Now, let's consider how we would use the box-and-whisker to find the central half of the data, or the interquartile range.

EXAMPLE

▶ Look again at Joy's Science Quiz Scores. What is the interquartile range of her quiz scores? What does this interquartile range mean?

▶ Look at the box section of the data display. Find the lowest score and the greatest score represented by the box. The lowest score represented by the box is 65, and the highest score represented by the box is 82.

▶ The interquartile range of Joy's scores on her science quizzes is 65 to 82. This means that half her quiz scores were between 65 and 82.

Scatter Plots and Trend Lines

A **scatter plot** is a graph of points that show the relationship between two sets of data. Each point is plotted according to its x-coordinate and its y-coordinate, but the points do not form a line. However, a general pattern, or **trend**, is often revealed in the placement of the points and, taken together, a **trend line** can be drawn using points that indicates the direction, or general tendency, of the data:

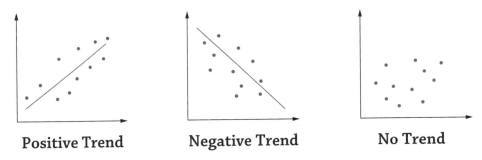

Positive Trend   **Negative Trend**   **No Trend**

The scatter plot on the next page shows the relationship between age and the number of hours of Internet usage per week.

Age and Internet Use

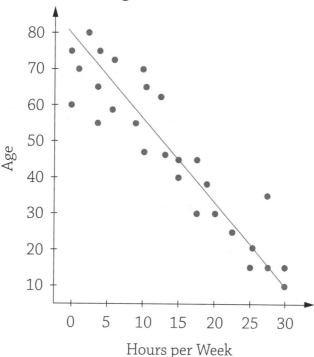

Age (vertical axis) vs. Hours per Week (horizontal axis)

▶ Refer to the Internet scatter plot above. How would you describe the trend line? What conclusion can you draw from the scatter plot?

▶ Study the values of the intervals on the horizontal axis and the vertical axis. The *x*-coordinates represent the number of hours of Internet use. The *x*-coordinates increase in value as you move from left to right. The *y*-coordinates show the age of the Internet users. The *y*-coordinates increase in value as you move from bottom to top.

▶ Look at the trend line that runs through the data points. The trend line is negative because it slopes down and to the right.

▶ Based on the intervals and negative trend line, it is reasonable to conclude that the older people are, the less time they use the Internet.

Misleading Graphs

As we've seen, graphs are used to present data in a way that we can readily understand it. Sometimes, though, graphs may not be constructed in the best way to present data accurately. When evaluating graphs, then, we should look for ways in which the display might distort the presentation of data. One common way to alter the appearance of data, especially in bar graphs, is to truncate (shorten) the intervals on the y-axis.

EXAMPLE

▶ The bar graph below intends to show customer satisfaction with trucks made by four different auto manufacturers.

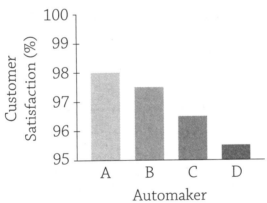

**Customer Satisfaction
for Trucks by Automaker**

▶ If you look quickly at the graph, it looks like customer satisfaction for trucks made by automaker A is twice what it is for trucks made by automaker C. Is this a reasonable conclusion? Why or why not?

▶ For automaker A, customer satisfaction is 98%, whereas, for automaker C, customer satisfaction is 96.5%.

▶ The difference in customer satisfaction between automaker A and automaker C is only 1.5% greater, not twice as great. The length of the bars creates a misleading impression about customer satisfaction because the y-axis starts at 95%.

A line graph may also misrepresent data trends by using inaccurate or inappropriate scales.

▶ What is misleading about the way data is presented in the line graph below? What causes the misrepresentation of the data?

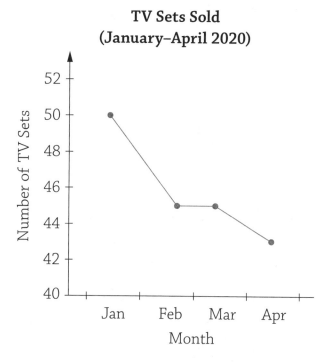

TV Sets Sold
(January–April 2020)

▶ For each data point, identify the number of TV sets sold on the vertical scale. The store sold 50 TV sets in January and 47 TV sets in February.

▶ The length of the line segment between these months is misleading because it seems to show that the store sold half as many TV sets in January as it did in February. The misconception happens beecause the scale on the *y*-axis starts at 40 and the intervals are very small.

BTW

Too many or too few intervals along the y-axis change the shape of a line graph, thereby misrepresenting the data.

EXERCISES

EXERCISE 12–1

A high school basketball team's top nine scores this season are listed in the data array shown below.

19, 20, 22, 23, 24, 27, 28, 28, 34

1. What is the mean of the scores?

2. What is the median of the scores?

3. What is the mode of the scores?

4. What is the range of the scores?

EXERCISE 12–2

The bar graph shows votes for favorite snacks by eighth-grade students at a local school.

1. Which snack was selected as the favorite and how many students voted for it?

2. Which two snacks received the same number of votes?

3. Which snack received the fewest votes?

4. Which snack received the second most number of votes?

EXERCISE 12-3

The histogram below shows the number of attendees at a concert by age group.

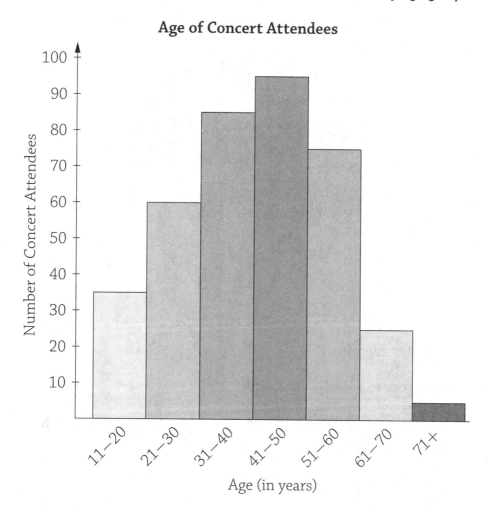

Age of Concert Attendees

1. Which age group had the most attendees at the concert?

2. Which age group had the fewest attendees at the concert?

3. Why is the age range 0–10 missing?

4. Based on the data, what generalization can be made about the overall age of concert attendees?

EXERCISE 12-4

The circle graph below shows students' favorite types of TV shows.

Favorite Types of TV Shows

Identify which TV show is:

1. the students' favorite.

2. the second most popular with students.

3. the students' least favorite.

4. twice as popular as drama.

EXERCISE 12–5

The line graph below shows the number of new cars sold last week by Ms. Specter at the auto dealership where she works.

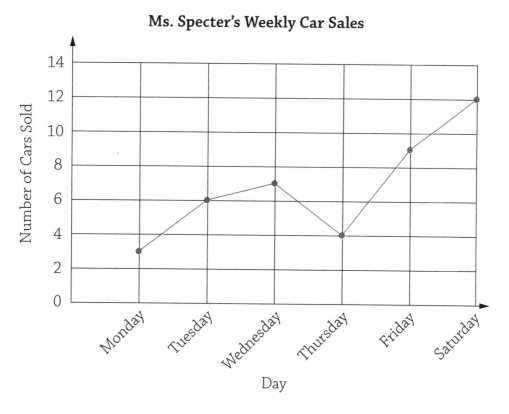

Ms. Specter's Weekly Car Sales

1. On which day did Ms. Specter sell the most number of cars?

2. On which day did Ms. Specter sell the fewest number of cars?

3. How many more cars did Ms. Specter sell on Wednesday than on Thursday?

4. How many cars in all did Ms. Specter sell last week?

EXERCISE 12–6

Use the stem-and-leaf plot that shows test scores for students on Mr. Rossi's final math test.

Final Math Test Scores, Mr. Rossi

Stem	Leaf
9	2 5
8	3 4 6 7 9
7	1 3 5 8 8 8
6	2 6 6 8
5	1 9

6 | 2 = 62 points

1. What is the lowest score on Mr. Rossi's final math test?

2. What is the highest score on Mr. Rossi's final math test?

3. What is the mode of scores on Mr. Rossi's final math test?

4. What is the range of scores on Mr. Rossi's final math test?

EXERCISE 12–7

Use the box-and-whisker plot that shows the weight in pounds of 13 pumpkins.

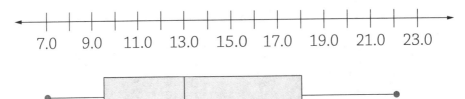

1. What is the median weight of these pumpkins?

2. What is the interquartile range of these pumpkins?

3. What is the weight of the pumpkin at the lower extreme?

4. What is the weight of the pumpkin at the upper extreme?

EXERCISE 12–8

Use the scatter plot that relates the height of a basketball player to the maximum number of free throws achieved in one game.

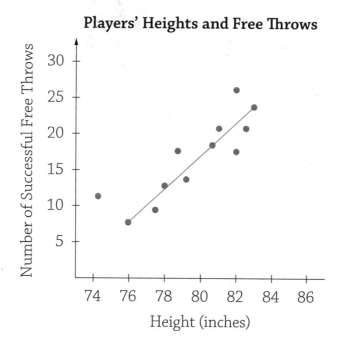

Players' Heights and Free Throws

1. What type of relationship does the trend line show between height and maximum number of free throws in a game?

2. What is a reasonable prediction of the number of free throws a player who is 86 inches tall will make? Why do you think your prediction is reasonable?

EXERCISE 12–9

Identify what is misleading about each data display.

1.

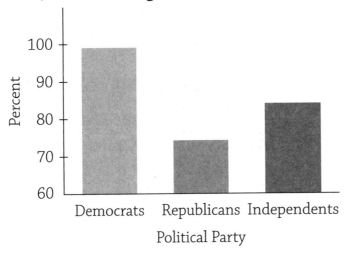

Paid Family and Medical Leave
(Percent Who Agree Based on Political Party)

2.

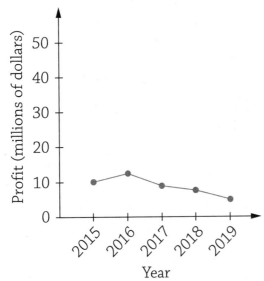

Capital Widgets
Profits, 2015–2019

Flashcard
App

Answer Key

1
Real Numbers

EXERCISE 1–1

1. The 3 is in the hundreds place and its value is three hundreds or 300.

2. The 5 is in the ten thousands place and its value is five ten thousands or 50,000.

3. The 4 is in the hundredths place and its value is four one hundredths or 0.04.

4. The 9 is in the thousandths place and its value is nine thousandths or 0.009.

EXERCISE 1–2

1. The absolute value of -3 is $|3|$.

2. The absolute value of 2 is $|2|$.

3. The absolute value of 4 is $|4|$.

4. The absolute value of -5 is $|5|$.

EXERCISE 1–3

1. rational

2. irrational

3. rational

4. irrational

EXERCISE 1–4

1. $-2 > -5$ or $-5 < -2$

2. $3 > -4$ or $-4 < 3$

3. $0.2 = \dfrac{1}{5}$ or $\dfrac{1}{5} = 0.2$

4. $\dfrac{6}{9} > 0.50$ or $0.50 < \dfrac{6}{9}$

EXERCISE 1–5

1. $-4 < -2 < \dfrac{-2}{5} < 0.25 < \dfrac{2}{3} < |3.5|$

2. $5.7 > 4 > 0.35 > \dfrac{-5}{8} > -5 > -5.2$

3. $-3.5 < -0.75 < \dfrac{-4}{7} < 3 < 3.2 < \sqrt{14}$

4. $\dfrac{-72}{-12} > \sqrt{35} > 5.8 > -0.45 > -\dfrac{9}{10} > -1.7$

EXERCISE 1–6

1. -7

2. -26

3. 9

4. -6

EXERCISE 1–7

1. 12
2. −2
3. 6
4. 0

EXERCISE 1–8

1. 24
2. −72
3. 0
4. −28

EXERCISE 1–9

1. −8
2. 6
3. −13
4. 9

EXERCISE 1–10

1. −56 minutes

   $-4 \times 14 = -56$

2. −$4 per share

   $-800 \div 200 = -4$

3. $3 per share

$$10 + (-8) + 6 + (-5)$$
$$= (10 + 6) + (-8) + (-5)$$
$$= 16 + (-8) + (-5)$$
$$= 16 + (-13)$$
$$= 3$$

4. The temperature dropped 26°F.

$$(15) - (-11) = 26$$

Fractions and Decimals

EXERCISE 2-1

1. $\dfrac{7}{10}$

2. $\dfrac{7}{12}$

3. $\dfrac{4}{9}$

4. $\dfrac{3}{7}$

EXERCISE 2-2

1. Possible answers: $\dfrac{6}{14}, \dfrac{9}{21}$

2. Possible answers: $\dfrac{10}{24}, \dfrac{15}{36}$

3. Possible answers: $\dfrac{22}{50}, \dfrac{44}{100}$

4. Possible answers: $\dfrac{46}{80}, \dfrac{69}{120}$

EXERCISE 2–3

1. $\dfrac{2}{3}$

2. $\dfrac{3}{5}$

3. $\dfrac{1}{4}$

4. $\dfrac{1}{3}$

EXERCISE 2–4

1. $\dfrac{12}{20} + \dfrac{6}{20} = \dfrac{18}{20} = \dfrac{9}{10}$

2. $\dfrac{13}{18} - \dfrac{7}{18} = \dfrac{6}{18} = \dfrac{1}{3}$

3. $\dfrac{3}{7} + \dfrac{5}{21} = \dfrac{9}{21} + \dfrac{5}{21} = \dfrac{14}{21} = \dfrac{2}{3}$

4. $\dfrac{9}{12} - \dfrac{1}{4} = \dfrac{9}{12} - \dfrac{3}{12} = \dfrac{6}{12} = \dfrac{1}{2}$

EXERCISE 2–5

1. $3\dfrac{1}{3} + 2\dfrac{3}{5} = \dfrac{10}{3} + \dfrac{13}{5} = \dfrac{50}{15} + \dfrac{39}{15} = \dfrac{89}{15} = 5\dfrac{14}{15}$

2. $7\dfrac{5}{9} - 2\dfrac{2}{3} = \dfrac{68}{9} - \dfrac{24}{9} = \dfrac{44}{9} = 4\dfrac{8}{9}$

3. $5\dfrac{2}{3} + 3\dfrac{4}{9} = \dfrac{17}{3} + \dfrac{31}{9} = \dfrac{51}{9} + \dfrac{31}{9} = \dfrac{82}{9} = 9\dfrac{1}{9}$

4. $8\dfrac{7}{10} - 2\dfrac{13}{15} = \dfrac{87}{10} - \dfrac{43}{15} = \dfrac{261}{30} - \dfrac{86}{30} = \dfrac{175}{30} = 5\dfrac{25}{30} = 5\dfrac{5}{6}$

EXERCISE 2–6

1. $\dfrac{3}{4} \times \dfrac{4}{5} = \dfrac{12}{20} = \dfrac{3}{5}$

2. $\dfrac{5}{8} \times \dfrac{3}{5} = \dfrac{15}{40} = \dfrac{3}{8}$

3. $3\dfrac{2}{3} \times 5\dfrac{3}{4} = \dfrac{11}{3} \times \dfrac{23}{4} = \dfrac{253}{12} = 21\dfrac{1}{12}$

4. $1\dfrac{2}{5} \times 4\dfrac{3}{7} = \dfrac{7}{5} \times \dfrac{31}{7} = \dfrac{217}{35} = 6\dfrac{1}{5}$

EXERCISE 2–7

1. $\dfrac{4}{9} \div \dfrac{2}{3} = \dfrac{4}{9} \times \dfrac{3}{2} = \dfrac{12}{18} = \dfrac{12 \div 6}{18 \div 6} = \dfrac{2}{3}$

2. $\dfrac{3}{4} \div \dfrac{1}{8} = \dfrac{3}{4} \times \dfrac{8}{1} = \dfrac{24}{4} = \dfrac{24 \div 4}{4 \div 4} = \dfrac{6}{1} = 6$

3. $2\dfrac{5}{6} \div \dfrac{1}{3} = \dfrac{17}{6} \div \dfrac{1}{3} = \dfrac{17}{6} \times \dfrac{3}{1} = \dfrac{51}{6} = 8\dfrac{3}{6} = 8\dfrac{1}{2}$

4. $3\dfrac{1}{2} \div 1\dfrac{2}{5} = \dfrac{7}{2} \div \dfrac{7}{5} = \dfrac{7}{2} \times \dfrac{5}{7} = \dfrac{35}{14} = \dfrac{35 \div 7}{14 \div 7} = \dfrac{5}{2} = 2\dfrac{1}{2}$

EXERCISE 2–8

1. $1\dfrac{4}{15}$ inches

$$\dfrac{2}{3}+\dfrac{3}{5}=\dfrac{10}{15}+\dfrac{9}{15}=\dfrac{19}{15}=1\dfrac{4}{15}$$

2. $1\dfrac{8}{15}$ ounces

$$6\dfrac{1}{5}-4\dfrac{2}{3}=\dfrac{31}{5}-\dfrac{14}{3}=\dfrac{93}{15}-\dfrac{70}{15}=\dfrac{23}{15}=1\dfrac{8}{15}$$

3. $5\dfrac{1}{4}$ cups

$$1\dfrac{1}{2}\times3\dfrac{1}{2}=\dfrac{3}{2}\times\dfrac{7}{2}=\dfrac{21}{4}=5\dfrac{1}{4}$$

4. 7 days

$$80\div10\dfrac{1}{2}=80\div\dfrac{21}{2}=\dfrac{80}{1}\times\dfrac{2}{21}=\dfrac{160}{21}=7\dfrac{13}{21}$$

EXERCISE 2–9

1. 0.52

2. 39.9

3. 241.125

4. 0.0015

EXERCISE 2–10

1. fifteen and two-tenths

2. twenty-one and forty-five hundredths

3. four thousandths

4. five and one hundred forty-one ten thousandths

EXERCISE 2–11

1. $>$

2. $=$

3. $<$

4. $>$

EXERCISE 2–12

1. 5.59

2. 9.705

3. 3.455

4. 5.84

EXERCISE 2–13

1. 9.8

2. 36.34

3. 39.6

4. 10.672

EXERCISE 2–14

1. $8.58
   $3.5 \times 2.45 = 8.575 \approx 8.58$

2. $0.35 or 35 cents
   $4.25 \div 12 = 0.3541\overline{66}$

3. $19.55
   $156.40 \div 8 = 19.55$

4. $32.48
   $2.5 \times 12.99 = 32.475 \approx 32.48$

3
Exponents, Roots, and Scientific Notation

EXERCISE 3-1

1. $3 \times 3 \times 3 = 3^3$
2. $2 \times 2 \times 2 \times 2 = 2^4$
3. $4 \times 4 \times 4 \times 4 \times 4 = 4^5$
4. $10 \times 10 \times 10 \times 10 \times 10 \times 10 = 10^6$

EXERCISE 3-2

1. $2^4 = 2 \times 2 \times 2 \times 2 = 16$
2. $5^3 = 5 \times 5 \times 5 = 125$
3. $8^1 = 8$
4. $10^6 = 10 \times 10 \times 10 \times 10 \times 10 \times 10 = 1{,}000{,}000$
5. $3^3 \times 100^0 = 27 \times 1 = 27$

EXERCISE 3-3

1. $2^{-3} = \dfrac{1}{2} \times \dfrac{1}{2} \times \dfrac{1}{2} = \dfrac{1}{8}$

2. $6^{-2} = \dfrac{1}{6} \times \dfrac{1}{6} = \dfrac{1}{36}$

3. $(-3)^5 = (-3) \times (-3) \times (-3) \times (-3) \times (-3) = -243$

4. $(-4)^4 = (-4) \times (-4) \times (-4) \times (-4) = 256$

EXERCISE 3–4

1. $5^2 \times 5^4 = 5^{2+4} = 5^6 = 15,625$
2. $2^5 \times 2^3 = 2^{5+3} = 2^8 = 256$
3. $3^4 \times 3^3 = 3^{4+3} = 3^7 = 2,187$
4. $7^2 \times 7^0 = 7^{2+0} = 7^2 = 49$

EXERCISE 3–5

1. $3^6 \div 3^2 = 3^{6-2} = 3^4 = 81$
2. $2^7 \div 2^3 = 2^{7-3} = 2^4 = 16$
3. $5^7 \div 5^2 = 5^{7-2} = 5^5 = 3,125$
4. $3^1 \div 10^0 = 3 \div 1 = 3$

EXERCISE 3–6

1. $(3^2)^3 = 3^{2\times3} = 3^6 = 729$
2. $(2^4)^2 = 2^{4\times2} = 2^8 = 256$
3. $(7^2)^2 = 7^{2\times2} = 7^4 = 2,401$
4. $(10^2)^4 = 10^{2\times4} = 10^8 = 100,000,000$

EXERCISE 3–7

1. $\sqrt{48} = \sqrt{16 \times 3} = 4\sqrt{3}$
2. $\sqrt{51}$ is a little more than $\sqrt{49}$, so $\sqrt{51} \approx 7.1$
3. $3.5\sqrt{6} + 7.8\sqrt{6} = 11.3\sqrt{6}$
4. $5.78\sqrt{3} - 6.26\sqrt{3} = -0.48\sqrt{3}$
5. $\sqrt{32} \times 2.9\sqrt{2} = 2.9 \times (\sqrt{32} \times \sqrt{2}) = 2.9 \times \sqrt{64} = 2.9 \times 8 = 23.2$
6. $12\sqrt{7} \div 3\sqrt{5} = \dfrac{12}{3} \div \sqrt{\dfrac{7}{5}} = 4\sqrt{\dfrac{7}{5}} = \dfrac{4\sqrt{7}}{\sqrt{5}} \times \dfrac{\sqrt{5}}{\sqrt{5}} = \dfrac{4\sqrt{35}}{5}$

EXERCISE 3–8

1. $2.4 \times 10^5 = 240{,}000$
2. $4.1 \times 10^8 = 410{,}000{,}000$
3. $0.00073 = 7.3 \times 10^{-4}$
4. $0.0000025 = 2.5 \times 10^{-6}$

EXERCISE 3–9

1. $5.91 \times 10^{11} = 591{,}000{,}000{,}000 \text{ m}$
2. $261{,}000{,}000 = 2.61 \times 10^8 \text{ km}$
3. $0.0085 = 8.5 \times 10^{-3} \text{ cm}$
4. $0.000007 = 7 \times 10^{-6} \text{ cm}$

Equations and Inequalities

EXERCISE 4–1

1. $w - 4$
2. $n \div 8$ or $\dfrac{n}{8}$
3. $x = 4 \times y$ or $x = 4y$
4. $a = 24 + z$

EXERCISE 4–2

1. $$\begin{aligned} w + 9 &= 12 \\ w + 9 - 9 &= 12 - 9 \\ w &= 3 \end{aligned}$$

2.
$$z - 4.75 = 10.5$$
$$z - 4.75 + 4.75 = 10.5 + 4.75$$
$$z = 15.25$$

3.
$$\frac{x}{4} = -6$$
$$\frac{4}{1} \times \frac{x}{4} = \frac{4}{1} \times -\frac{6}{1}$$
$$\frac{4}{4}x = -24$$
$$x = -24$$

4.
$$\frac{2.5}{2.5}y = \frac{40}{2.5}$$
$$y = 16$$

EXERCISE 4–3

1.

$$z + 9 \geq 12$$
$$z + 9 - 9 \geq 12 - 9$$
$$z \geq 3$$

2.

$$5x - 14 < 26$$
$$5x < 40$$
$$\frac{5x}{5} < \frac{40}{5}$$
$$x < 8$$

3.

Number line from −9 to 9 with an open circle at 6.

$$y - 4 > 2$$
$$y - 4 + 4 > 2 + 4$$
$$y > 6$$

4.

Number line from −9 to 9 with an open circle at −6.

$$\frac{-1}{3}y + 5 < 7$$

$$\frac{-1}{3}y + 5 - 5 < 7 - 5$$

$$\frac{-1}{3}y < 2$$

$$\frac{3}{1} \times \frac{-1y}{3} < \frac{3}{1} \times \frac{2}{1}$$

$$\frac{-3y}{3} < \frac{6}{1}$$

$$-y < 6$$

$$\frac{-y}{-1} < \frac{6}{-1}$$

$$y > -6$$

EXERCISE 4–4

1. The party mix cost $58.

$$c = s + b + p$$
$$85 = 15 + 12 + p$$
$$85 = 27 + p$$
$$85 - 27 = 27 - 27 + p$$
$$58 = p$$

2. 245 points

$$\frac{x}{5} = 49$$

$$\frac{5}{1} \times \frac{x}{5} = \frac{49}{1} \times \frac{5}{1}$$

$$\frac{5x}{5} = \frac{245}{1}$$

$$x = 245$$

3. $365f = 4{,}500{,}000{,}000$

$$\frac{365}{365}f = \frac{4{,}500{,}000{,}000}{365}$$

$$f = 12{,}328{,}767$$

$$f \approx 12{,}329{,}000$$

4. Ari must make at least 19 one-way trips for the monthly pass to cost less than one-way trips.

$$3x > 54$$

$$\frac{3x}{3} = \frac{54}{3}$$

$$x > 18$$

5. Briana can buy 15 plants and pay the $5 delivery charge for $50.

$$3p + 5 \leq 50$$

$$3p + 5 - 5 \leq 50 - 5$$

$$3p \leq 45$$

$$\frac{3p}{3} \leq \frac{45}{3}$$

$$p \leq 15$$

Ratio, Proportion, and Percent

EXERCISE 5–1

1. 5 to 9, 5:9, $\dfrac{5}{9}$

2. $\dfrac{46}{52} = \dfrac{12}{13}$

3. $\dfrac{96}{162} = \dfrac{16}{27}$

4. $\dfrac{18}{42} = \dfrac{3}{7}$

EXERCISE 5–2

1. Possible answers: 6:10 and 9:15

2. Yes, $\dfrac{16}{20}$ and $\dfrac{48}{60}$ are equivalent ratios.

$$\dfrac{16}{20} = \dfrac{16 \div 4}{20 \div 4} = \dfrac{4}{5} \text{ and } \dfrac{48}{60} = \dfrac{48 \div 12}{60 \div 12} = \dfrac{4}{5}$$

EXERCISE 5–3

1. $=$

2. $<$

3. $>$

4. $=$

EXERCISE 5–4

1. Raul should use 8 ounces of red paint and 12 ounces of yellow paint to make 20 ounces of orange paint. To solve, find equivalent ratios.

$$\frac{2}{3} = \frac{2 \times 4}{3 \times 4} = \frac{8}{12}$$

2. Bruce's typing rate is 37 words per minute.

$$\frac{185 \text{ words}}{5 \text{ minutes}} = \frac{37 \text{ words}}{1 \text{ minute}}$$

3. Elena has a greater ratio of red to blue marbles since $\frac{3}{5} > \frac{1}{2}$.

$$\text{Bret} = \frac{6}{12} = \frac{1}{2} \text{ and Elena} = \frac{9}{15} = \frac{3}{5}$$

4. 32 miles per gallon

$$\frac{576 \text{ miles}}{18 \text{ gallons}} = 32 \text{ miles per gallon}$$

EXERCISE 5–5

1. 12

$$\frac{9}{x} = \frac{3}{4}$$
$$3x = 36$$
$$\frac{3x}{3} = \frac{36}{3}$$
$$x = 12$$

2. 40

$$\frac{18}{12} = \frac{60}{x}$$
$$18x = 720$$
$$\frac{18x}{18} = \frac{720}{18}$$
$$x = 40$$

3. 325 miles

$$\frac{195}{3} = \frac{x}{5}$$

$$3x = 975$$

$$\frac{3x}{3} = \frac{975}{3}$$

$$x = 325$$

4. $39.00

$$\frac{26}{24} = \frac{x}{36}$$

$$24x = 936$$

$$\frac{24x}{24} = \frac{936}{24}$$

$$x = 39$$

EXERCISE 5–6

1. 0.68 or 68%

2. $\frac{35}{100}$ and 0.35

3. 90

$$
\begin{array}{r}
1500 \\
\times\ 0.06 \\
\hline
9000 \\
000 \\
\underline{000} \\
90.00
\end{array}
$$

4. 4 is 5% of 80.

$$80 \times x\% = 4$$

$$\frac{80x}{80} = \frac{4}{80}$$

$$x = 0.05$$

EXERCISE 5-7

1. 30%

$$169 - 130 = 39$$

$$\rightarrow \frac{\text{amount of increase}}{\text{original number}} = \frac{39}{130} = 0.3$$

$$0.3 = 30\%$$

2. 16.7%

$$13,884 - 11,570 = 2,314$$

$$\rightarrow \frac{\text{amount of decrease}}{\text{original number}} = \frac{2,314}{13,884} \approx 0.1\overline{666}$$

$$0.1666 \approx 16.7\%$$

3. $1,000

$$I = 5,000 \times 0.05 \times 4 = 1,000$$

4. $1,800

$$I = 10,000 \times 0.06 \times 3 = 1,800$$

Plane Geometry

EXERCISE 6–1

1. $\overline{RS}$
2. $\overleftrightarrow{PQ}$
3. $\overleftrightarrow{LO}$
4. $\overrightarrow{TU}$

EXERCISE 6–2

1. acute angle
2. obtuse angle
3. right angle
4. straight angle

EXERCISE 6–3

1. obtuse triangle
2. acute triangle
3. right triangle

EXERCISE 6–4

1. isosceles triangle
2. scalene triangle
3. equilateral triangle

EXERCISE 6-5

1. 13 cm

$$a^2 + b^2 = c^2$$
$$5^2 + 12^2 = c^2$$
$$25 + 144 = c^2$$
$$169 = c^2$$
$$c = 13$$

2. 12 m

$$a^2 + b^2 = c^2$$
$$9^2 + b^2 = 15^2$$
$$81 + b^2 = 225$$
$$81 - 81 + b^2 = 225 - 81$$
$$b^2 = 144$$
$$b = 12$$

3. $7\sqrt{2}$ in   This is a 45°–45°–90° triangle, so the length of the hypotenuse is the length of one side times $\sqrt{2}$.

4. 4 cm   This is a 30°–60°–90° triangle, so the length of the shorter leg is half the length of the hypotenuse.

EXERCISE 6-6

1. $\angle A = 180° - 117° = 63°$
2. $\angle A = 180° - 47° = 133°$
3. $\angle 3 = \angle 7 = 72°$
4. $\angle 4 = \angle 8 = 108°$

EXERCISE 6–7

1. $\angle x = 90° - 65° = 25°$
2. $\angle x = 180° - 105° = 75°$
3. $\angle x = 180° - 86° = 94°$
4. $\angle x = 180° - 147° = 33°$

EXERCISE 6–8

1. $900°$

   $(n - 2)180° = (7 - 2)180° = (5)180° = 900°$

2. $1,260°$

   $(n - 2)180° = (9 - 2)180° = (7)180° = 1,260°$

3. $120°$

   $$\frac{(n - 2)180°}{8} = \frac{(8 - 2)180°}{8} = \frac{(6)180°}{8} = \frac{960°}{8} = 120°$$

4. $144°$

   $$\frac{(n - 2)180°}{10} = \frac{(10 - 2)180°}{10} = \frac{(8)180°}{10} = \frac{1440°}{10} = 144°$$

EXERCISE 6–9

1. 35 m

   $5\text{ m} + 10\text{ m} + 5\text{ m} + 15\text{ m} = 35\text{ m}$

2. 15 cm

   $5\text{ cm} + 4\text{ cm} + 6\text{ cm} = 15\text{ cm}$

3. 80 ft^2

   $A = bh$
   $ = (8\text{ ft} \times 16\text{ ft})$
   $ = 128\text{ ft}^2$

4. 30 in^2

$$A = \frac{1}{2}bh$$

$$= \frac{1}{2}(5 \text{ in} \times 12 \text{ in})$$

$$= \frac{1}{2}(60 \text{ in}^2)$$

$$= 30 \text{ in}^2$$

EXERCISE 6–10

1. 31.4 m
$$C = 2\pi r$$
$$= 2 \times 5 \times 3.14$$
$$= 31.4 \text{ m}$$

2. 25.1 in
$$C = \pi d$$
$$= 8 \times 3.14$$
$$= 25.1 \text{ in}$$

3. 28.3 ft^2
$$A = \pi r^2$$
$$\approx 3.14 \times (3^2)$$
$$\approx 3.14 \times (9)$$
$$\approx 28.3 \text{ ft}^2$$

4. 50.2 cm^2
$$A = \pi r^2$$
$$\approx 3.14 \times (4^2)$$
$$\approx 3.14 \times (16)$$
$$\approx 50.24 \text{ cm}^2$$

7

Solid Geometry

EXERCISE 7–1

1. 944 in^2 Use the formula for the surface area of a rectangular prism: SA(rectangular prism) $= 2lw + 2lh + 2wh$. Notice that the length $=$ 16 inches, its width $= 12$ inches, and its height $= 10$ inches.

SA(top/bottom of box) $= 2lw = 2 \times 16$ in $\times 12$ in $= 384$ in^2

SA(long sides of box) $= 2lh = 2 \times 16$ in $\times 10$ in $= 320$ in^2

SA(short sides of box) $= 2wh = 2 \times 12$ in $\times 10$ in $= 240$ in^2

SA(box) $= 384$ in$^2 + 320$ in$^2 + 240$ in$^2 = 944$ in^2

2. 132 ft^2 Use the formula for the surface area of a rectangular prism: SA(rectangular prism) $= 2lw + 2lh + 2wh$. Notice that the length of the platform $= 8$ feet, its width $= 5$ feet, and its height $= 2$ feet.

SA(top/bottom of platform) $= 2lw = 2 \times 8$ ft $\times 5$ ft $= 80$ ft^2

SA(long sides of platform) $= 2lh = 2 \times 8$ ft $\times 2$ ft $= 32$ ft^2

SA(short sides of platform) $= 2wh = 2 \times 5$ ft $\times 2$ ft $= 20$ ft^2

SA(platform) $= 80$ ft$^2 + 32$ ft$^2 + 20$ ft$^2 = 132$ ft^2

EXERCISE 7–2

1. 175.8 cm^2 Use the formula for the surface area of a cylinder: SA (cylinder) $= 2\pi rh + 2\pi r^2$. Notice that the diameter of the can/cylinder is 8 centimeters, so its radius is 4 centimeters, and its height $=$ 3 centimeters.

SA(can face) $= 2\pi rh = 2 \times 3.14 \times 4$ cm $\times 3$ cm $= 75.36$ cm^2

SA(can bases) $= 2\pi r^2 = 2 \times 3.14 \times 4$ cm $\times 4$ cm $= 100.48$ cm^2

SA(can) $= 75.36$ cm$^2 + 100.48$ cm$^2 \approx 175.8$ cm^2

2. 88 in^2 Use the formula for the surface area of a cylinder: SA (cylinder) $= 2\pi rh + 2\pi r^2$. Notice that the radius $= 2$ inches and the height $= 5$ inches.

$SA(\text{candle face}) = 2\pi rh = 2 \times 3.14 \times 2 \text{ in} \times 5 \text{ in} = 62.8 \text{ in}^2$

$SA(\text{candle bases}) = 2\pi r^2 = 2 \times 3.14 \times 2 \text{ in} \times 2 \text{ in} = 25.12 \text{ in}^2$

$SA(\text{candle}) = 62.8 \text{ in}^2 + 25.12 \text{ in}^2 = 87.92 \text{ in}^2 \approx 88 \text{ in}^2$

EXERCISE 7–3

1. 138 cm^2 Use the formula for the surface area of a cone: SA (cone) $= \pi rs + \pi r^2$. Notice that the diameter of the base of paperweight is 8 centimeters, so its radius $= 4$ centimeter, and its slant height $= 7$ centimeters.

$SA(\text{paperweight face}) = \pi rs = 3.14 \times 4 \text{ cm} \times 7 \text{ cm} = 87.92 \text{ cm}^2$

$SA(\text{paperweight base}) = \pi r^2 = 3.14 \times 4 \text{ cm} \times 4 \text{ cm} = 50.24 \text{ cm}^2$

$SA(\text{paperweight}) = 87.92 \text{ cm}^2 + 50.24 \text{ cm}^2 = 138.16 \text{ cm}^2 \approx 138 \text{ cm}^2$

2. $1{,}243 \text{ in}^2$ Use the formula for the surface area of a cone: $SA = \pi rs + \pi r^2$. Notice that the radius of the base of sculpture $= 12$ inches and its slant height $= 21$ inches.

$SA(\text{scultpture face}) = \pi rs = 3.14 \times 12 \text{ in}^2 \times 21 \text{ in}^2 = 791.28 \text{ in}^2$

$SA(\text{scultpture base}) = \pi r^2 = 3.14 \times 12 \text{ in}^2 \times 12 \text{ in}^2 = 452.16 \text{ in}^2$

$SA(\text{scultpture}) = 791.28 \text{ in}^2 + 452.16 \text{ in}^2 = 1{,}243.44 \text{ in}^2 \approx 1{,}243 \text{ in}^2$

EXERCISE 7–4

1. 95 cm^2 Use the formula for the surface area of a pyramid: $SA = B(\text{area}) + 4F(\text{face area})$. Notice that the base of the paperclip pyramid is a square with each side of the base $= 5$ centimeters and the slant height $= 7$ centimeters.

$SA(\text{Base}) = 5 \text{ cm} \times 5 \text{ cm} = 25 \text{ cm}^2$

$SA(\text{Face}) = \dfrac{1}{2}(7 \text{ cm} \times 5 \text{ cm}) = \dfrac{1}{2}(35) = 17.5 \text{ cm}^2$

$SA(\text{Sum of Face Areas}) = 4 \times 17.5 \text{ cm}^2 = 70 \text{ cm}^2$

$SA(\text{pyramid}): 25 \text{ cm}^2 + 70 \text{ cm}^2 = 95 \text{ cm}^2$

2. **528 ft²**   Use the formula for the surface area of a pyramid: $SA = B(\text{area}) + 4F(\text{face area})$. Notice that the base of the pyramid is a square with each side of the base = 12 feet and the slant height = 16 feet.

$SA(\text{Base}) = 12 \text{ ft} \times 12 \text{ ft} = 144 \text{ ft}^2$

$SA(\text{Face}) = \dfrac{1}{2}(16 \text{ ft} \times 12 \text{ ft}) = \dfrac{1}{2}(192) = 96 \text{ ft}^2$

$SA(\text{Sum of Face Areas}) = 4 \times 96 \text{ ft}^2 = 384 \text{ ft}^2$

$SA(\text{pyramid}): 144 \text{ ft}^2 + 384 \text{ ft}^2 = 528 \text{ ft}^2$

EXERCISE 7–5

1. **1,520 cm²**   Use the formula for the surface area of a sphere: $SA = 4\pi r^2$. Notice that the diameter of the bowling ball is 22 centimeters, so its radius is 11 centimeters.

$SA = 4 \times 3.14 \times 11\text{cm} \times 11 \text{ cm} = 1{,}519.76 \text{ cm}^2 \approx 1{,}520 \text{ cm}^2$

2. **1,017 cm²** Use the formula for the surface area of a sphere: $SA = 4\pi r^2$. Notice that the radius is 11 centimeters.

$SA = 4 \times 3.14 \times 9 \text{ cm} \times 9 \text{ cm} = 1{,}017.36 \text{ cm}^2 \approx 1{,}017 \text{ cm}^2$

EXERCISE 7–6

1. **60 ft³**   Use the formula for the volume of a rectangular prism: $V = lwh$.

$V = 6 \text{ ft} \times 5 \text{ ft} \times 2 \text{ ft} = 60 \text{ ft}^3$

2. **3,000 in³**   Use the formula for the volume of a rectangular prism: $V = lwh$.

$V = 10 \text{ in} \times 10 \text{ in} \times 30 \text{ in} = 3{,}000 \text{ in}^3$

EXERCISE 7–7

1. 38 in³   Use the formula for the volume of a cylinder: $V = \pi r^2 h$. Notice that the diameter of the cylinder is 4 inches, so its radius $= 2$ inches and its height $= 3$ inches.

$$V = 3.14 \times 2 \text{ in} \times 2 \text{ in} \times 3 \text{ in} = 37.69 \text{ in}^3 \approx 38 \text{ in}^3$$

2. 2,826 m³   Use the formula for the volume of a cylinder: $V = \pi r^2 h$. Notice that the water storage tank's diameter is 30 meters, so its radius $= 15$ meters and its height $= 4$ meters.

$$V = 3.14 \times 15 \text{ m} \times 15 \text{ m} \times 4 \text{ m} = 2,826 \text{ m}^3$$

EXERCISE 7–8

1. 377 in³   Use the formula for the volume of a cone: $V = \dfrac{1}{3} \pi r^2 h$.

Notice that the cone's radius $= 6$ inches and its height $= 10$ inches.

$$V = \frac{1}{3}(3.14 \times 6 \text{ in} \times 6 \text{ in} \times 10 \text{ in})$$

$$= \frac{1}{3}(1,130.04 \text{ in}^3) = 376.8 \text{ in}^3 \approx 377 \text{ in}^3$$

2. 81,771 mm³   Use the formula for the volume of a cone: $V = \dfrac{1}{3} \pi r^2 h$.

Notice that the diameter of the cone is 50 millimeters, so its radius $=$ 25 millimeters and its height $= 125$ millimeters.

$$V = \frac{1}{3}(3.14 \times 25 \text{ mm} \times 25 \text{ mm} \times 125 \text{ mm})$$

$$= \frac{1}{3}(245,312 \text{ mm}^3) = 81,770.8\overline{33} \text{ mm}^3 \approx 81,771 \text{ mm}^3$$

EXERCISE 7–9

1. 467 in^3 Use the formula for the volume of a square pyramid: $V = \dfrac{1}{3}Bh$.

Notice that the pyramid has a square base with each side $= 10$ inches and its height $= 14$ inches.

$$B = (10 \text{ in})^2 = 100 \text{ in}^2$$

$$V = \frac{1}{3}Bh$$

$$= \frac{1}{3}(100 \text{ in}^2 \times 14 \text{ in})$$

$$= \frac{1}{3}(1{,}400 \text{ in}^3) = 466.\overline{66} \text{ in}^3 \approx 467 \text{ in}^3$$

2. 256 cm^3 Use the formula for the volume of a square pyramid:

$V = \dfrac{1}{3}Bh$. Notice that the pyramid has a square base with each side $=$

8 centimeters and its height $= 12$ centimeters.

$$B = (8 \text{ cm} \times 8 \text{ cm}) = 64 \text{ cm}^2$$

$$V = \frac{1}{3}Bh$$

$$= \frac{1}{3}(64 \text{ cm}^2 \times 12 \text{ cm})$$

$$= \frac{1}{3}(768 \text{ cm}^3) = 256 \text{ cm}^3$$

EXERCISE 7–10

1. $5{,}572 \text{ cm}^3$ Use the formula for the volume of a sphere: $V \text{ (sphere)} =$

$\dfrac{4}{3}\pi r^3$. The radius of the soccer ball $= 11$ centimeters.

$$V = \frac{4}{3}(3.14 \times 11 \text{ cm} \times 11 \text{ cm} \times 11 \text{ cm})$$

$$= \frac{4}{3}(4{,}179.34 \text{ cm}^3) \approx 5572.45\overline{33} \text{ cm}^3 \approx 5{,}572 \text{ cm}^3$$

2. $4{,}187$ mm^3 Use the formula for the volume of a sphere: V (sphere) $= \frac{4}{3}\pi r^3$. Notice that the diameter of the sphere is 20 millimeters, so its radius $= 10$ millimeters.

$$V = \frac{4}{3}(3.14 \times 10 \text{ mm} \times 10 \text{ mm} \times 10 \text{ mm})$$

$$= \frac{4}{3}(3{,}140 \text{ mm}^3) \approx 4186.\overline{66} \text{ mm}^3 \approx 4{,}187 \text{ mm}^3$$

Coordinate Geometry

EXERCISE 8–1

1. Point A has coordinates $(2, 2)$ in Quadrant I.
   Point B has coordinates $(-4, -3)$ in Quadrant III.
   Point C has coordinates $(1, -5)$ in Quadrant IV.
   Point D has coordinates $(-4, 3)$ in Quadrant II.

2. The coordinates $(0, -4)$ locate point E.
   The coordinates $(1, 5)$ locate point F.
   The coordinates $(-4, 0)$ locate point G.
   The coordinates $(-5, -2)$ locate point H.

EXERCISE 8–2

1. Length of line segment $\overline{BC}$: 4 in

   The coordinates of point B are $(2, 2)$, and the coordinates of point C are $(6, 2)$.

$$d = \sqrt{(x_2 - x_1)^2 + (y_2 - y_1)^2}$$
$$d(\overline{BC}) = \sqrt{(6 - 2)^2 + (2 - 2)^2}$$
$$= \sqrt{(4)^2 + (0)^2}$$
$$= \sqrt{16}$$
$$= 4 \text{ in}$$

2. Length of line segment $\overline{AB}$: 3 in

   The coordinates of point A are (2, 5), and the coordinates of point B are (2, 2).

$$d = \sqrt{(x_2 - x_1)^2 + (y_2 - y_1)^2}$$
$$d(\overline{AB}) = \sqrt{(2 - 2)^2 + (2 - 5)^2}$$
$$= \sqrt{(0)^2 + (-3)^2}$$
$$= \sqrt{9}$$
$$= 3 \text{ in}$$

3. Length of line segment $\overline{AC}$: 5 in

   The coordinates of point A are (2, 5), and the coordinates of point C are (6, 2).

$$d = \sqrt{(x_2 - x_1)^2 + (y_2 - y_1)^2}$$
$$d(\overline{AC}) = \sqrt{(6 - 2)^2 + (2 - 5)^2}$$
$$= \sqrt{(4)^2 + (-3)^2}$$
$$= \sqrt{16 + 9}$$
$$= \sqrt{25}$$
$$= 5 \text{ in}$$

4. Perimeter of $\triangle ABC = 4 \text{ in} + 3 \text{ in} + 5 \text{ in} = 12 \text{ in}$

5. Area of $\triangle ABC = \dfrac{1}{2} bh = \dfrac{1}{2} (4 \times 3) = \dfrac{1}{2} \times 12 = 6 \text{ in}^2$

EXERCISE 8–3

1. reflection
2. translation
3. tessellation
4. rotation

EXERCISE 8–4

1. reflection over the x-axis
2. quadrant II
3. quadrant III
4. Check the coordinates of each point on the diagram or switch coordinates and multiply the new x-coordinate by -1.

   $A'(-4, -5)$, $B'(-5, -4)$, $C'(-4, -2)$, $D'(-3, -4)$

EXERCISE 8–5

1. 90° rotation, clockwise
2. quadrant I
3. quadrant IV
4. Check the coordinates of each point on the diagram or switch coordinates and multiply the new y-coordinate by -1.

   $P'(2, -2)$, $Q'(4, -3)$, $R'(4, -6)$, $S'(2, -6)$

EXERCISE 8–6

1. translation, 1 unit right and 6 units up
2. quadrant IV
3. quadrant I
4. Check the coordinates of each point on the diagram.

   $W'(3, 4)$, $X'(6, 4)$, $Y'(5, 2)$, $Z'(2, 2)$

EXERCISE 8–7

1. 6 lines of symmetry

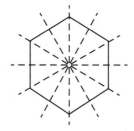

2. 2 lines of symmetry

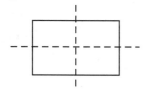

3. 5 lines of symmetry

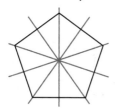

4. 0 lines of symmetry

EXERCISE 8–8

1. Semi-regular tessellation; squares and equilateral triangles
2. Semi-regular tessellation; irregular octagon and regular parallelogram

Linear Functions

EXERCISE 9–1

1. Not a function; The input value $x = 4$ has two outputs values $y = 5$ and $y = 8$.

2. Function; Each input value x has only one output value.

3. Function; Even though each output value is 4; nonetheless, each input value x has only one output value.

4. Not a function; The input value $x = 5$ has two outputs values $y = 7$ and $y = -7$.

EXERCISE 9–2

1. $y = x - 2$, or the value of y equals the value of x minus 2.

2. $y = x + 5$, or the value of y equals the value of x plus 5.

3. $y = x - 3$, or the value of y equals the value of x minus 3.

4. $y = x + 4$, or the value of y equals the value of x plus 4.

EXERCISE 9–3

1. $y = 2x + 2$

Input (x)	−2	−1	0	1	2
Output (y)	−2	0	2	4	6

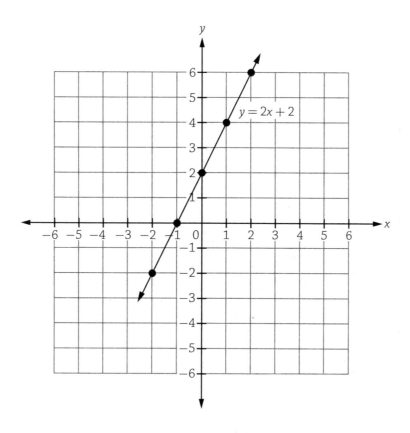

$$y = 2x + 2$$

2. $y = 2x - 2$

Input (x)	−2	−1	0	1	2
Output (y)	−6	−4	−2	0	2

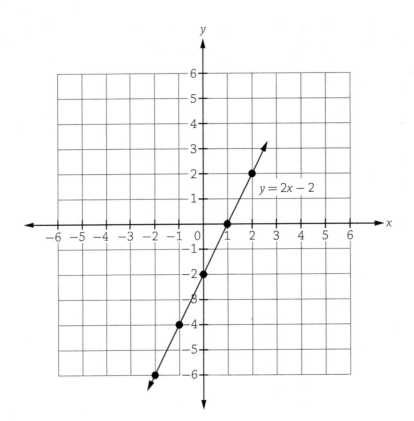

EXERCISE 9–4

1. x-intercept $= (-2, 0)$
   y-intercept $= (0, 3)$

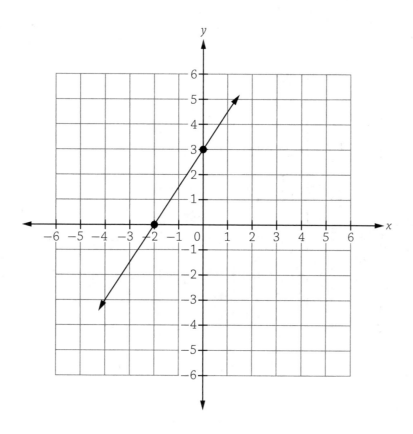

2. *x*-intercept $= (4, 0)$
   y-intercept $= (0, 2)$

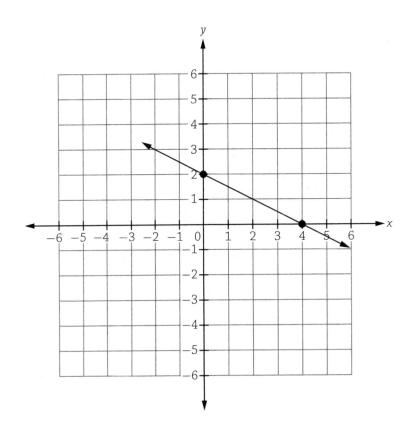

EXERCISE 9-5

1. 3

   (2, 1) and (3, 4)

   $$\text{slope} = \frac{rise}{run} = \frac{y_2 - y_1}{x_2 - x_1}$$

   $$= \frac{4 - 1}{3 - 2} = \frac{3}{1} = 3$$

2. -4

$(2, 6)$ and $(3, 2)$

$$\text{slope} = \frac{rise}{run} = \frac{y_2 - y_1}{x_2 - x_1}$$

$$= \frac{2 - 6}{3 - 2} = \frac{-4}{1} = -4$$

3. 2

$(4, -3)$ and $(7, 3)$

$$\text{slope} = \frac{rise}{run} = \frac{y_2 - y_1}{x_2 - x_1}$$

$$= \frac{3 - (-3)}{7 - 4} = \frac{6}{3} = 2$$

4. -5

$(1, 4)$ and $(3, -6)$

$$\text{slope} = \frac{rise}{run} = \frac{y_2 - y_1}{x_2 - x_1}$$

$$= \frac{-6 - 4}{3 - 1} = \frac{-10}{2} = -5$$

EXERCISE 9-6

1. Slope $= 2$; y-intercept $= 4$

Equation: $y = 2x + 4$

$$\text{slope} = \frac{rise}{run} = \frac{y_2 - y_1}{x_2 - x_1}$$

$$= \frac{-2 - 2}{-3 - (-1)} = \frac{-4}{-2} = 2$$

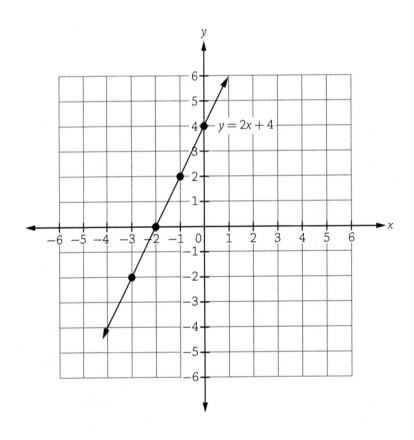

2. Slope $= -3$; y-intercept $= 1$

   Equation: $y = -3x + 1$

   $$\text{slope} = \frac{rise}{run} = \frac{y_2 - y_1}{x_2 - x_1}$$

   $$= \frac{-5 - 4}{2 - (-1)} = \frac{-9}{3} = -3$$

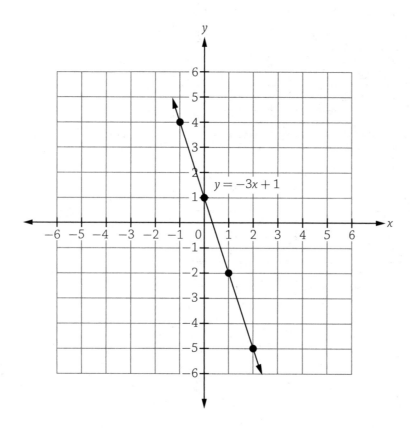

$y = -3x + 1$

EXERCISE 9–7

1. (0, 2), independent

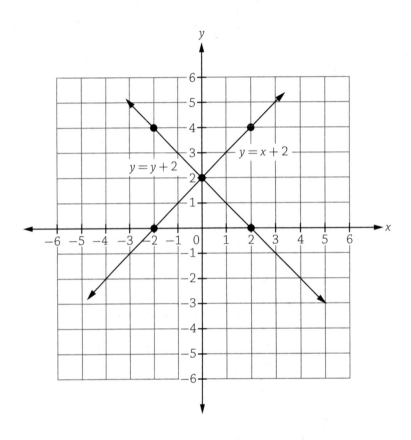

2. Infinite, dependent

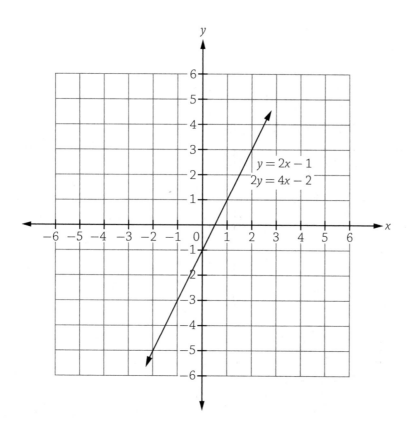

$y = 2x - 1$
$2y = 4x - 2$

3. (−1, 4), independent

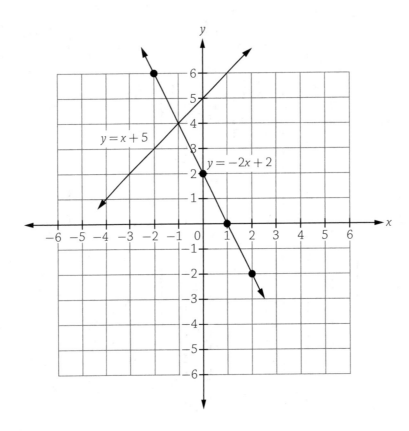

4. No solution, inconsistent

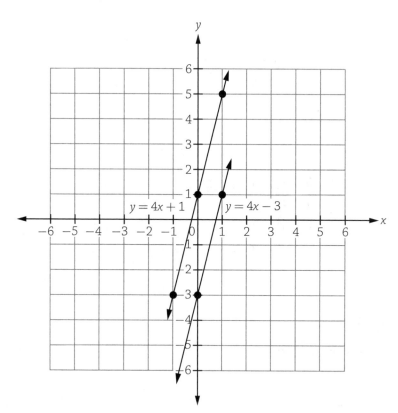

EXERCISE 9–8

1. $(-3, 2)$

$x = y - 5$ (Equation 1)
$2x + y = -4$ (Equation 2)

Solve for y in Equation 2 by substituting $y - 5$ for x.

$$2x + y = -4$$
$$2(y - 5) + y = -4$$
$$2y - 10 + y = -4$$
$$3y - 10 = -4$$

$$3y - 10 + 10 = -4 + 10$$
$$3y = 6$$
$$y = 2$$

Solve for x in Equation 1 by substituting 2 for y.

$$x = y - 5$$
$$x = 2 - 5$$
$$x = -3$$

Check:

$x = y - 5$	$2x + y = -4$
$-3 = 2 - 5$	$2(-3) + 2 = -4$
$-3 = -3$	$-6 + 2 = -4$
	$-4 = -4$

2. $(-4, 2)$

$$x + 4y = 4 \text{ (Equation 1)}$$
$$x - y = -6 \text{ (Equation 2)}$$

Find the value of x in Equation 1.

$$x + 4y = 4$$
$$x + 4y - 4y = 4 - 4y$$
$$x = -4y + 4$$

Solve for y in Equation 2 by substituting $-4y + 4$ for x.

$$x - y = -6$$
$$(-4y + 4) - y = -6$$
$$-5y + 4 = -6$$
$$-5y + 4 - 4 = -6 - 4$$
$$-5y = -10$$
$$\frac{-5y}{-5} = \frac{-10}{-5}$$
$$y = 2$$

Solve for x in Equation 1 by substituting 2 for y.

$$x + 4y = 4$$
$$x + 4(2) = 4$$
$$x + 8 = 4$$
$$x + 8 - 8 = 4 - 8$$
$$x = -4$$

Check:

$$x + 4y = 4 \qquad x - y = -6$$
$$-4 + 4(2) = 4 \qquad -4 - 2 = -6$$
$$-4 + 8 = 4 \qquad -6 = -6$$
$$4 = 4$$

3. $(3, 3)$

$$y + 2x = 9 \text{ (Equation 1)}$$
$$y = 2x - 3 \text{ (Equation 2)}$$

Solve for x in Equation 1 by substituting $2x - 3$ for y.

$$y + 2x = 9$$
$$(2x - 3) + 2x = 9$$
$$4x - 3 = 9$$
$$4x - 3 + 3 = 9 + 3$$
$$4x = 12$$
$$x = 3$$

Solve for y in Equation 2 by substituting 3 for x.

$$y = 2x - 3$$
$$y = 2(3) - 3$$
$$y = 6 - 3$$
$$y = 3$$

Check:

$$\begin{aligned} y + 2x &= 9 \\ 3 + 2(3) &= 9 \\ 3 + 6 &= 9 \\ 9 &= 9 \end{aligned} \qquad \begin{aligned} y &= 2x - 3 \\ 3 &= 2(3) - 3 \\ 3 &= 6 - 3 \\ 3 &= 3 \end{aligned}$$

4. $(-1, -2)$

$y = 6x + 4$ (Equation 1)
$2y = -6x - 10$ (Equation 2)

Solve for x by substituting $6x + 4$ for y in Equation 2.

$$\begin{aligned} 2y &= -6x - 10 \\ 2(6x + 4) &= -6x - 10 \\ 12x + 8 &= -6x - 10 \\ 12x + 6x + 8 &= -6x + 6x - 10 \\ 18x + 8 &= -10 \\ 18x + 8 - 8 &= -10 - 8 \\ 18x &= -18 \\ x &= -1 \end{aligned}$$

Solve for y by substituting -1 for x in Equation 1.

$y = 6x + 4$
$y = 6(-1) + 4$
$y = -6 + 4$
$y = -2$

Check:

$$\begin{aligned} y &= 6x + 4 \\ -2 &= 6(-1) + 4 \\ -2 &= -6 + 4 \\ -2 &= -2 \end{aligned} \qquad \begin{aligned} 2y &= -6x - 10 \\ 2(-2) &= -6(-1) - 10 \\ -4 &= 6 - 10 \\ -4 &= -4 \end{aligned}$$

EXERCISE 9–9

1. Write the inequality as an equation in $y = mx + b$ form:

$$y - 2x > 3 \quad \rightarrow \quad y - 2x = 3 \quad \rightarrow \quad y = 2x + 3$$

Graph the inequality. Use a dashed line since the inequality sign is $<$. Use $(0, 0)$ as a test point and shade the half plane that contains the solutions.

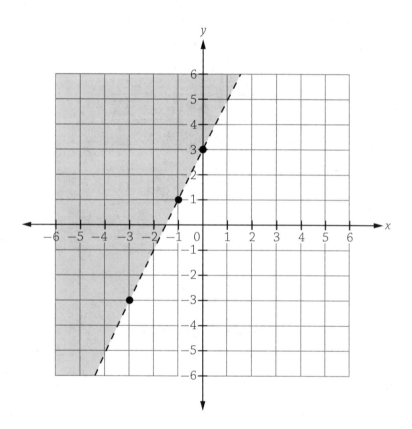

2. Write the inequality as an equation in $y = mx + b$ form:

$$y \leq -2x + 3 \quad \rightarrow \quad y = -2x + 3$$

Graph the inequality. Use a solid line since the inequality sign is $\leq$. Use $(0, 0)$ as a test point and shade the half plane that contains the solutions.

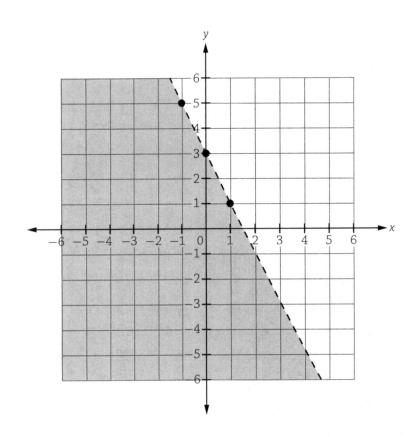

10
Polynomials

EXERCISE 10-1

1. $14x^3 - 4x - 6$, third degree

   $6x^3 - 4x + 8x^3 - 6$
   $= (6x^3 + 8x^3) - 4x - 6$
   $= 14x^3 - 4x - 6$

2. $5t^2 + 4t + 16$, second degree

$12 + 7t^2 + 4 - 2t^2 + 4t$
$= (7t^2 - 2t^2) + 4t + (12 + 4)$
$= 5t^2 + 4t + 16$

3. $10n^2 - 8n - 2$, second degree

$-2 + 4n^2 - 8n + 6n^2$
$= (4n^2 + 6n^2) - 8n - 2$
$= 10n^2 - 8n - 2$

4. $c^4 + 2c^2 + 9c + 8$, fourth degree

$9c + 7c^4 + 10 - 6c^4 + 2c^2 - 2$
$= (7c^4 - 6c^4) + 2c^2 + 9c + (10 - 2)$
$= c^4 + 2c^2 + 9c + 8$

EXERCISE 10–2

1.
$$\begin{array}{r} 4y^2 - 2y + 6 \\ + \ 2y^2 + 5y - 2 \\ \hline 6y^2 + 3y + 4 \end{array}$$

2.
$$\begin{array}{r} 5z^2 - 4z + 3 \\ + \ 3z^2 \qquad + 1 \\ \hline 8z^2 - 4z + 4 \end{array}$$

3.
$$\begin{array}{r} 2s^2 + 7s + 5 \\ + \ 4s^2 - 6s - 3 \\ \hline 6s^2 + \ s + 2 \end{array}$$

4.
$$\begin{array}{r} 3x^2 - 3x + 4 \\ + \ 9x^2 \qquad - 8 \\ \hline 12x^2 - 3x - 4 \end{array}$$

EXERCISE 10–3

1.
$$\begin{array}{r} 8w^2 - 4 \\ + \ -(2w^2 + 3) \\ \hline \end{array}$$
$\rightarrow$
$$\begin{array}{r} 8w^2 - 4 \\ + \ -2w^2 - 3 \\ \hline 6w^2 - 7 \end{array}$$

2.
$$\begin{array}{r} 5a^2 - 6a + 8 \\ + \ -(4a^2 - a + 6) \\ \hline \end{array}$$
$\rightarrow$
$$\begin{array}{r} 5a^2 - 6a + 8 \\ + -4a^2 + a - 6 \\ \hline a^2 - 5a + 2 \end{array}$$

3.
$$\begin{array}{r} 9t^2 + 5t - 7 \\ + \ -(5t^2 - 4t) \\ \hline \end{array}$$
$\rightarrow$
$$\begin{array}{r} 9t^2 + 5t - 7 \\ + \ -5t^2 + 4t \\ \hline 4t^2 + 9t - 7 \end{array}$$

4.
$$\begin{array}{r} -4x^2 + 8x + 5 \\ + \ -(2x^2 - 4x - 3) \\ \hline \end{array}$$
$\rightarrow$
$$\begin{array}{r} -4x^2 + 8x + 5 \\ + \ -2x^2 + 4x - 3 \\ \hline -6x^2 + 12x + 8 \end{array}$$

EXERCISE 10–4

1. $5x(4x^2 - 2)$
   $= (5x \times 4x^2) + (5x \times -2)$
   $= 20x^3 - 10x$

2. $3y(-2y^2 + 3)$
   $= (3y \times -2y^2) + (3y \times 3)$
   $= -6y^3 + 9y$

3. $(5s^3)^2$
   $= (5^2) \times (s^3)^2$
   $= 25s^6$

4. $4t(5t^2 + 4)$
   $= (4t \times 5t^2) + (4t \times 4)$
   $= 20t^3 + 16t$

EXERCISE 10–5

1. $(x + 6)(2x - 2)$

First terms: $x \times 2x = 2x^2$
Outer terms: $x \times -2 = -2x$
Inner terms: $6 \times 2x = 12x$
Last terms: $6 \times -2 = -12$

$2x^2 - 2x + 12x - 12 \rightarrow 2x^2 + 10x - 12$

2. $(n^2 - 4)(n^2 + 2)$

First terms: $n^2 \times n^2 = n^4$
Outer terms: $n^2 \times 2 = 2n^2$
Inner terms: $-4 \times n^2 = -4n^2$
Last terms: $-4 \times 2 = -8$

$n^4 + 2n^2 - 4n^2 - 8 \rightarrow n^4 - 2n^2 - 8$

3. $(2w - 5)(3w + 7)$

First terms: $2w \times 3w = 6w^2$
Outer terms: $2w \times 7 = 14w$
Inner terms: $-5 \times 3w = -15w$
Last terms: $-5 \times 7 = -35$

$6w^2 + 14w - 15w - 35 \rightarrow 6w^2 - w - 35$

4. $(5a^2 - 3)(-2a + 3)$

First terms: $5a^2 \cdot -2a = -10a^3$
Outer terms: $5a^2 \cdot 3 = 15a^2$
Inner terms: $-3 \cdot -2a = 6a$
Last terms: $-3 \cdot 3 = -9$

$-10a^3 + 15a^2 + 6a - 9$

EXERCISE 10–6

1. $\dfrac{8y^2 - 10y + 6}{2y}$

$$= \dfrac{8y^2}{2y} + \dfrac{-10y}{2y} + \dfrac{6}{2y}$$

$$= 4y - 5 + \dfrac{3}{y}$$

2. $\dfrac{-9x^2 + 15x - 12}{3x}$

$$= \dfrac{9x^2}{3x} + \dfrac{15x}{3x} - \dfrac{12}{3x}$$

$$= 3x + 5 - \dfrac{4}{x}$$

3. $\dfrac{12t^2 - 16t + 8}{4t}$

$$= \dfrac{12t^2}{4t} - \dfrac{16t}{4t} + \dfrac{8}{4t}$$

$$= 3t - 4 + \dfrac{2}{t}$$

4.
$$\frac{-20w^2 + 12w + 8}{2w}$$

$$= \frac{-20w^2}{2w} + \frac{12w}{2w} + \frac{8}{2w}$$

$$= -10w + 6 + \frac{4}{w}$$

11
Probability

EXERCISE 11-1

1. There are 24 permutations: $4 \times 3 \times 2 = 24$. As a list with 1, 2, 3, 4 representing the CDs:

1, 2, 3, 4	2, 1, 3, 4	3, 1, 2, 4	4, 1, 2, 3
1, 2, 4, 3	2, 1, 4, 3	3, 1, 4, 2	4, 1, 3, 2
1, 3, 2, 4	2, 3, 1, 4	3, 2, 1, 4	4, 2, 1, 3
1, 3, 4, 2	2, 3, 4, 1	3, 2, 4, 1	4, 2, 3, 1
1, 4, 2, 3	2, 4, 1, 3	3, 4, 1, 2	4, 3, 1, 2
1, 4, 3, 2	2, 4, 3, 1	3, 4, 2, 1	4, 3, 2, 1

2. There are 120 possible permutations in which the teams can place first, second, and third: $6 \times 5 \times 4 = 120$.

EXERCISE 11-2

1. There are 10 possible combinations of tacos and drinks.

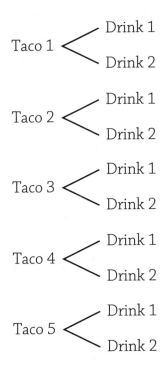

2. There are 6 permutations of the three balls.

1, 2, 3
1, 3, 2
2, 1, 3
2, 3, 1
3, 1, 2
3, 2, 1

EXERCISE 11–3

1. The probability that you will spin a number greater than 6 is $\dfrac{1}{4}$.

   Only the numbers 7 and 8 are greater than 6.

   $$\rightarrow P = \frac{2}{8} = \frac{1}{4}$$

2. The probability that you will select a blue chip is $0.1\overline{66}$.

$$P = \frac{5}{30} = \frac{1}{6}$$

$$\rightarrow \frac{1}{6} = 0.1\overline{66}$$

3. The probability that you will draw a vowel is 40%.

$$P = \frac{2}{5}$$

$$\rightarrow \frac{2}{5} = 0.40 = 40\%$$

4. The probability of selecting a card with an even number is $\frac{6}{11}$.

There are 5 odd numbers and 6 even numbers in 20 through 30.

$$\rightarrow P = \frac{6}{11}$$

EXERCISE 11–4

1. The probability that a coin lands heads up and the number rolled is less than 3 is $\frac{1}{6}$.

$$P(\text{heads}) = \frac{1}{2}$$

$$P(\text{less than 3}) = \frac{2}{6} = \frac{1}{3}$$

$$P(\text{heads, less than 3}) = \frac{1}{2} \times \frac{1}{3} = \frac{1}{6}$$

2. The probability that both tiles will be greater than 5 is 0.22.

$$P(\text{greater than 5, first draw}) = \frac{5}{10} = \frac{1}{2}$$

$$P(\text{greater than 5, second draw}) = \frac{4}{9}$$

$$P(\text{greater than 5, both draws}) = \frac{1}{2} \times \frac{4}{9} = \frac{4}{8} = \frac{2}{9} = 0.22$$

3. The probability that you will select two blue counters is $\frac{2}{15}$.

$$P(\text{blue}_1) = \frac{4}{10} = \frac{2}{5}$$

$$P(\text{blue}_2) = \frac{3}{9} = \frac{1}{3}$$

$$P(\text{blue}_1 \text{ and blue}_2) = \frac{4}{10} \times \frac{3}{9} = \frac{12}{90} = \frac{2}{15}$$

4. The probability that you will select two Fuji apples 0.147.

$$P(\text{Fuji, first draw}) = \frac{8}{20} = \frac{2}{5}$$

$$P(\text{Fuji, second draw}) = \frac{7}{19}$$

$$P(\text{Fuji, both draws}) = \frac{2}{5} \times \frac{7}{19} = \frac{14}{95} \approx 0.147$$

EXERCISE 11–5

1. The probability that the next randomly chosen student will select science fiction as his or her favorite type of movie is 0.28.

$$P = \frac{55}{200} = 0.275 \approx 0.28$$

2. The probability that the next time Jeni spins, the spinner will land on 1 is 20%.

$$P = \frac{6}{30} = 0.20 = 20\%$$

EXERCISE 11-6

1. The probability that a point chose at random on JK falls on JL is $\frac{1}{4}$.

$$\frac{\text{Length of } \overline{JL}}{\text{Length of } \overline{JK}} = \frac{2}{8} = \frac{1}{4}$$

2. The probability that a dart will land in the shaded area is 21.5%.

$$P = \frac{\text{shaded area}}{\text{total area}}$$

$$= \frac{A(\text{square}) - A(\text{circle})}{A(\text{square})}$$

$$= \frac{s^2 - \pi r^2}{s^2}$$

$$= \frac{4^2 - \pi(2)^2}{4^2}$$

$$= \frac{16 - 4\pi}{16}$$

$$= \frac{16 - 4(3.14)}{16}$$

$$= \frac{16 - 12.56}{16}$$

$$= \frac{3.44}{16}$$

$$= 0.215 = 21.5\%$$

12
Data and Statistics

EXERCISE 12-1

1. 25

   $19 + 20 + 22 + 23 + 24 + 27 + 28 + 28 + 34 = 225; 225 \div 9 = 25$

2. 24

   Since the data set has 9 items, the fifth item is the median.

3. 28

   The number 28 occurs most frequently.

4. 15

   $34 - 19 = 15$

EXERCISE 12-2

1. Fruit was voted the favorite snack by 25 students.
2. Trail mix and Vegetables both received 15 votes.
3. Cheese and Crackers received only the fewest votes, only 12.
4. Popcorn received 23 votes, the second highest.

EXERCISE 12-3

1. The 41–50 age group had the most number of attendees at the concert.
2. The 71+ age group had the fewest number of attendees at the concert.
3. Sample answer: No one in that age range attended the concert.
4. Sample answer: The vast majority of concert attendees were between 31 and 60 years old.

EXERCISE 12–4

1. Action TV shows are the students' favorite at 30%.

2. Sci-FI TV shows are the second most popular with students at 25%.

3. Drama TV shows are the students' least favorite at 10%.

4. Comedy TV shows at 20% are twice as popular as dramas at 10%.

EXERCISE 12–5

1. On Saturday, Ms. Specter sold 12 cars.

2. On Monday, Ms. Specter sold 3 cars.

3. Ms. Specter sold 3 more cars on Wednesday than she did on Thursday: $7 - 4 = 3$.

4. Ms. Specter sold 41 cars in all. $(3 + 6 + 7 + 4 + 9 + 12 = 41)$

EXERCISE 12–6

1. The lowest score is 51.

2. The highest score is 95.

3. The mode is 78.

4. The range of scores is 44: $95 - 51 = 44$.

EXERCISE 12–7

1. The median weight is 13 pounds.

2. The interquartile range is between 9.5 pounds and 18.0 pounds.

3. The weight of the pumpkin at the lower extreme is 7.0 pounds.

4. The weight of the pumpkin at the upper extreme is 22.0 pounds.

EXERCISE 12–8

1. The relationship is positive. The taller the player, the more free throws he is successful with.

2. The player will probably make about 30 free throws. If you extend the trend line, you can see that the y-axis is at 30 when the x-axis is at 86.

EXERCISE 12–9

1. The vertical scale starts at 60, so it exaggerates the differences in the height of the bars. If the vertical scale started at 0 and increased regular at intervals of 10, the difference in heights would be much less. In fact, the data shows 94%, 74%, and 83%. Yet, the bars make it appear that the value for Republicans is more than half that of Democrats.

2. The vertical axis has too many intervals for the data and does not inidcate that the numbers are percentages. All the changes in the line occur between the 0 and 10. The number of intervals flattens the line and makes the changes seem less significant. In fact, profits fell 50% from 2015 and 2019.

NOTES

NOTES